Handbook of
ORGANIC FARMING
Principles and Practices

Handbook of ORGANIC FARMING
Principles and Practices

Rajat Kumar

Oxford Book Company
Jaipur ▪ New Delhi

Published By

Oxford Book Company

Regd. Off. "Bony Residency", Gate No. 2
Opp. Tilak Public School, Vishweswariya Nagar,
Gopalpura Road, Jaipur - 302018, India
Ph: 0141 - 2761280, Telefax: 2761381
email - oxfordbookjpr@gmail.com
website - oxfordbookcompany.com

Delhi Office: 102, 1st Floor, Satyam House,
4327/3, Ansari Road, Darya Ganj,
New Delhi - 110002, India
Ph: 011-45652440

Distributed By
KEYA INTERNATIONAL PRESS LLP
E-mail: keyainternationalpressllp@gmail.com

Published in India, 2025

ISBN : 9789350304921

Typeset by Shivangi Computers, Jaipur

Printed and bound in India by Thomson Press Ltd.,
New Delhi

Contents

Preface

Organic food proponents express concern over the potential negative effects of various chemical cultivation methods and genetic modification techniques used in modern conventional agriculture. The effect of pesticide residues from crop spraying, the presence of veterinary drugs in meat products, and the relatively unknown long term impacts of genetically modified varieties and breeds are all encompassed. Organic food is seen as avoiding relieving these concerns by prohibiting such practices. At present, there are no definitive scientific conclusions on any of these matters; individual studies are cited on both sides of the debate.

Organic food proponents cite the existence of reduced levels of pesticides and herbicides as a way to reduce the long term risk of chemical consumption. The degree of risk posed by pesticide residues remains uncertain. Pesticide use in conventional food products is heavily regulated, with established, research-based maximum residue levels (MRLs) below which residues are considered safe for human consumption. Also, many pesticides are not cumulative in the body, and are regularly eliminated. Notable exceptions include heavy metals such as lead or mercury which are sometimes found in foodstuffs in countries which have lax food production standards. The U.S. and most of Europe prohibit the use of inorganic compounds containing heavy metals in any type of agriculture including conventional. Organic farming standards do not allow the use of synthetic pesticides, but they do allow the use of certain so-called natural pesticides, such as those derived from plants. The most common organic pesticides, accepted for restricted use by most organic standards, include Bt, pyrethrum, and rotenone. Some organic pesticides, such as rotenone, have high toxicity to fish and aquatic creatures with some toxicity to mammals including humans.

A study published by the National Research Council in 1993 determined that for infants and children, the major source of exposure to pesticides is through diet. A recent study in 2006 measured the levels of organophosphorus pesticide exposure in 23 school children before and after replacing their diet with organic food. In this study it was found that levels of organophosphorus pesticide exposure dropped dramatically and immediately when the children switched to an organic diet.

The volumes capture the entire panorama of traditional agriculture, practiced through different cultures, across the globe. It would prove very useful to the researchers, teachers, students and environmentalists engaged in eco friendly sustainable agriculture.

— ***Rajat Kumar***

Chapter 1

Introduction

Organic farming is the form of agriculture that relies on techniques such as crop rotation, green manure, compost and biological pest control to maintain soil productivity and control pests on a farm. Organic farming uses fertilizers and pesticides but excludes or strictly limits the use of manufactured (synthetic) fertilizers, pesticides (which include herbicides, insecticides and fungicides), plant growth regulators such as hormones, livestock antibiotics, food additives, and genetically modified organisms.

Organic agricultural methods are internationally regulated and legally enforced by many nations, based in large part on the standards set by the International Federation of Organic Agriculture Movements (IFOAM), an international umbrella organisation for organic farming organisations established in 1972. IFOAM defines the overarching goal of organic farming as:

> *"Organic agriculture is a production system that sustains the health of soils, ecosystems and people. It relies on ecological processes, biodiversity and cycles adapted to local conditions, rather than the use of inputs with adverse effects. Organic agriculture combines tradition, innovation and science to benefit the shared environment and promote fair relationships and a good quality of life for all involved.."*
>
> —*International Federation of Organic Agriculture Movements*

Since 1990, the market for organic products has grown from nothing, reaching $55 billion in 2009 according to Organic Monitor. This demand has driven a similar increase in organically managed farmland. Approximately 37,000,000 hectares (91,000,000 acres) worldwide are now farmed organically, representing approximately 0.9 percent of total world farmland (2009).

History

Organic farming (of many particular kinds) was the original type of agriculture, and has been practiced for thousands of years. After the industrial revolution had introduced inorganic methods, some of which were not well developed and had serious side effects, an organic movement began in the 1940s as a reaction to agriculture's growing reliance on synthetic fertilizers.

Artificial fertilizers had been created during the 18th century, initially with superphosphates and then ammonia-based fertilizers mass-produced using the Haber-Bosch process developed during World War I. These early fertilizers were cheap, powerful, and easy to transport in bulk. Similar advances occurred in chemical pesticides in the 1940s, leading to the decade being referred to as the 'pesticide era'.

Although organic farming is prehistoric in the widest sense, Sir Albert Howard is widely considered to be the "father of organic farming" in the sense that he was a key founder of the post-industrial-revolution organic movement. Further work was done by J.I. Rodale in the United States, Lady Eve Balfour in the United Kingdom, and many others across the world. The modern organic movement is a revival movement in the sense that it seeks to restore balance that was lost when technology grew rapidly in the 19th and 20th centuries.

Modern organic farming has made up only a fraction of total agricultural output from its beginning until today. Increasing environmental awareness in the general population has transformed the originally supply-driven movement to a demand-driven one. Premium prices and some government subsidies attracted farmers. In the developing world, many producers farm according to traditional methods which are comparable to organic farming but are not certified.

In other cases, farmers in the developing world have converted for economic reasons.Organic farming (of many particular kinds in different eras and places) was the original type of agriculture, and has been practiced for thousands of years. After the industrial revolution had introduced inorganic methods, some of which were not well developed and had serious side effects, an organic movement began in the 1940s as a reaction to agriculture's growing reliance on synthetic fertilizers. The history of this modern revival of organic farming dates back to the first half of the 20th century at a time when there was a growing reliance on synthetic fertilizers.

Pre-World War II

The first 40 years of the 20th century saw simultaneous advances in biochemistry and engineering that rapidly and profoundly changed farming. The introduction of the gasoline-powered internal combustion engine ushered in the era of the tractor and made possible hundreds of mechanized farm implements. Research in plant breeding led to the commercialisation of hybrid seed. And a new manufacturing process made nitrogen fertilizer — first synthesized in the mid-19th century — affordably abundant. These factors changed the labour equation: there were some 600 tractors in the US around 1910 and over 3,000,000 by 1950; in 1900, it took one farmer to feed 2.5 people. Currently the ratio is 1 to well over 100. Fields grew bigger and cropping more specialised to make more efficient use of machinery.

Consciously organic agriculture (as opposed to the agriculture of indigenous cultures, which always employs only organic means) began more or less simultaneously in Central Europe and India. The British botanist Sir Albert Howard is often referred to as the father of modern organic agriculture. From 1905 to 1924, he worked as an agricultural adviser in Pusa, Bengal, where he documented traditional Indian farming practices and came to regard them as superior to his conventional agriculture science. His research and further development of these methods is recorded in his writings, notably, his 1940 book, *An Agricultural Testament*, which influenced many scientists and farmers of the day.

In Germany, Rudolf Steiner's development, biodynamic agriculture, was probably the first comprehensive organic farming system. This began with a lecture series Steiner presented at a farm in Koberwitz (now in Poland) in 1924. Steiner emphasized the farmer's role in guiding and balancing the interaction of the animals, plants and soil. Healthy animals depended upon healthy plants (for their food), healthy plants upon healthy soil, healthy soil upon healthy animals (for the manure).

In 1909, American agronomist F.H. King toured China, Korea, and Japan, studying traditional fertilization, tillage, and general farming practices. He published his findings in *Farmers of Forty Centuries* (1911, Courier Dover Publications, ISBN 0-486-43609-8). King foresaw a "world movement for the introduction of new and improved methods" of agriculture and in later years his book became an important organic reference.

The term *organic farming* was coined by Lord Northbourne in his book *Look to the Land* (written in 1939, published 1940). From his conception of "the farm as organism," he described a holistic, ecologically balanced approach to farming.

In 1939, influenced by Sir Albert Howard's work, Lady Eve Balfour launched the Haughley Experiment on farmland in England. It was the first scientific, side-by-side comparison of organic and conventional farming. Four years later, she published *The Living Soil*, based on the initial findings of the Haughley Experiment. Widely read, it led to the formation of a key international organic advocacy group, the Soil Association.

In Japan, Masanobu Fukuoka, a microbiologist working in soil science and plant pathology, began to doubt the modern agricultural movement. In 1937, he quit his job as a research scientist, returned to his family's farm in 1938, and devoted the next 60 years to developing a radical no-till organic method for growing grain and many other crops, now known as Nature Farming, 'do–nothing' farming or Fukuoka farming.

Post-World War II

Technological advances during World War II accelerated post-war innovation in all aspects of agriculture, resulting in large advances in mechanization (including large-scale irrigation), fertilization, and pesticides. In particular, two chemicals that had been produced in quantity for warfare, were repurposed for peace-time agricultural uses. Ammonium nitrate, used in munitions, became an abundantly cheap source of nitrogen. And a range of new pesticides appeared: DDT, which had been used to control disease-carrying insects around troops, became a general insecticide, launching the era of widespread pesticide use.

At the same time, increasingly powerful and sophisticated farm machinery allowed a single farmer to work larger areas of land and fields grew bigger.

In 1944, an international campaign called the Green Revolution was launched in Mexico with private funding from the US. It encouraged the development of hybrid plants, chemical controls, large-scale irrigation, and heavy mechanization in agriculture around the world.

During the 1950s, sustainable agriculture was a topic of scientific interest, but research tended to concentrate on developing the new chemical approaches. In the US, J.I. Rodale began to popularise the term and methods of organic growing, particularly to consumers through promotion of organic gardening.

In 1962, Rachel Carson, a prominent scientist and naturalist, published *Silent Spring*, chronicling the effects of DDT and other pesticides on the environment. A bestseller in many countries, including the US, and widely read around the world, *Silent Spring* is widely considered as being a key factor in the US government's 1972 banning of DDT. The book and its author are often credited with launching the worldwide environmental movement.

In the 1970s, global movements concerned with pollution and the environment increased their focus on organic farming. As the distinction between organic and conventional food became clearer, one goal of the organic movement was to encourage consumption of locally grown food, which was promoted through slogans like "Know Your Farmer, Know Your Food".

In 1972, the International Federation of Organic Agriculture Movements (IFOAM) was founded in Versailles, France and dedicated to the diffusion and exchange of information on the principles and practices of organic agriculture of all schools and across national and linguistic boundaries.

In 1975, Fukuoka released his first book, *The One-Straw Revolution*, with a strong impact in certain areas of the agricultural world. His approach to small-scale grain production emphasized a meticulous balance of the local farming ecosystem, and a minimum of human interference and labour.

In the 1980s, around the world, farming and consumer groups began seriously pressuring for government regulation of organic production. This led to legislation and certification standards being enacted through the 1990s and to date.

Since the early 1990s, the retail market for organic farming in developed economies has been growing by about 20% annually due to increasing consumer demand. Concern for the quality and safety of food, and the potential for environmental damage from conventional agriculture, are apparently responsible for this trend.

Twenty-first Century

Throughout this history, the focus of agricultural research and the majority of publicized scientific findings has been on chemical, not organic, farming. This emphasis has continued to biotechnologies like genetic engineering. One recent survey of the UK's leading government funding agency for bioscience research and training indicated 26 GM crop projects, and only one related to organic agriculture. This imbalance is largely driven by agribusiness in general, which, through research funding and government lobbying, continues to have a predominating effect on agriculture-related science and policy.

Agribusiness is also changing the rules of the organic market. The rise of organic farming was driven by small, independent producers and by consumers. In recent years, explosive organic market growth has encouraged the participation of agribusiness interests. As the volume and variety of "organic" products increases, the viability of the small-scale organic farm is at risk, and the meaning of organic farming as an agricultural method is ever more easily confused with the related but separate areas of organic food and organic certification.

In Havana, Cuba, a unique situation has made organic food production a necessity. Since the collapse of the Soviet Union in 1989 and its economic support, Cuba has had to produce food in creative ways like instituting the world's only state-supported infrastructure to support urban food production. Called organopónicos, the city is able to provide an ever increasing amount of its produce organically.

Methods

> *"An organic farm, properly speaking, is not one that uses certain methods and substances and avoids others; it is a farm whose structure is formed in imitation of the structure of a natural system that has the integrity, the independence and the benign dependence of an organism" —Wendell Berry, "The Gift of Good Land"*

Organic farming methods combine scientific knowledge of ecology and modern technology with traditional farming practices based on naturally occurring biological processes. Organic farming methods are studied in the field of agroecology. While conventional agriculture uses synthetic pesticides and water-soluble synthetically purified fertilizers, organic farmers are restricted by regulations to using natural pesticides and fertilizers. The principal methods of organic farming include crop rotation, green manures and compost, biological pest control, and mechanical cultivation. These measures use the natural environment to enhance agricultural productivity: legumes are planted to fix nitrogen into the soil, natural insect predators are encouraged, crops are rotated to confuse pests and renew soil, and natural materials such as potassium bicarbonate and mulches are used to control disease and weeds. Organic farmers are careful in their selection of plant breeds, and organic researchers produce hardier plants through plant breeding rather than genetic engineering.

Crop Diversity

Crop diversity is a distinctive characteristic of organic farming. Conventional farming focuses on mass production of one crop in one location, a practice called monoculture. This makes apparent economic sense: the larger the growing area, the lower the per unit cost of fertilizer, pesticides and specialised machinery for a single plant species. The science of agroecology has revealed the benefits of polyculture (multiple crops in the same space), which is often employed in organic farming. Planting a variety of vegetable crops supports a wider range of beneficial insects, soil microorganisms, and other factors that add up to overall farm health, but managing the balance requires expertise and close attention.

Farm Size

Farm size in great measure determines the general approach and specific tools and methods. Today, major food corporations are involved in all aspects of organic production on a large scale. However, organic farming originated as a small-scale enterprise, with operations from under 1-acre (4,000 m^2) to under 100 acres (0.40 km^2). The mixed vegetable organic market garden is often associated with fresh, locally grown produce, farmers' markets and the like, and this type of farm is often under 10 acres (40,000 m^2). Farming at this scale is generally labour-intensive, involving more manual labour and less mechanization. The type of crop also determines size: organic grain farms often involve much larger area. Larger organic farms tend to use methods and equipment similar to conventional farms, centred around the tractor.

Plant Nutrition

Soil Fertility

The central farming activity of fertilization illustrates the differences. Organic farming relies heavily on the natural breakdown of organic matter, using techniques like green manure and composting, to replace nutrients taken from the soil by previous crops. This biological process, driven by microorganisms such as mycorrhiza, allows the natural production of nutrients in the soil throughout the growing season, and has been referred to as *feeding the soil to feed the plant.* In chemical farming, individual nutrients, like nitrogen, are synthesized in a more or less pure form that plants can use immediately, and applied on a man-made schedule. Each nutrient is defined and addressed separately. Problems that may arise from one action (e.g. too much nitrogen left in the soil) are usually addressed with additional, corrective products and procedures (e.g. using water to wash excess nitrogen out of the soil).

Organic farming uses a variety of methods to improve soil fertility, including crop rotation, cover cropping, reduced tillage, and application of compost. By reducing tillage, soil is not inverted and exposed to air; less carbon is lost to the atmosphere resulting in more soil organic carbon. This has an added benefit of carbon sequestration which can reduce green house gases and aid in reversing climate change.

Pest Control

Different approaches to pest control are equally notable. In chemical farming, a specific insecticide may be applied to quickly kill off a particular insect pest (animal). Chemical controls can dramatically reduce pest populations for the short term, yet by unavoidably killing (or starving) natural predator insects and animals, cause an ultimate increase in the pest population. Repeated use of insecticides and herbicides and other pesticides also encourages rapid natural selection of resistant insects, plants and other organisms, necessitating increased use, or requiring new, more powerful controls.

In contrast, organic farming tends to tolerate some pest populations while taking a longer-term approach. Organic pest control involves the cumulative effect of many techniques, including:

- allowing for an acceptable level of pest damage;
- encouraging predatory beneficial insects to control pests;
- encouraging beneficial microorganisms and insects; this by serving them nursery plants and/or an alternative habitat, usually in a form of a shelterbelt, hedgerow, or beetle bank
- careful crop selection, choosing disease-resistant varieties
- planting companion crops that discourage or divert pests;
- using row covers to protect crops during pest migration periods;
- using pest regulating plants and biologic pesticides and herbicides
- using no-till farming, and no-till farming techniques as false seedbeds
- rotating crops to different locations from year to year to interrupt pest reproduction cycles;
- Using insect traps to monitor and control insect populations.

Each of these techniques also provides other benefits—soil protection and improvement, fertilization, pollination, water conservation, season extension, etc.—and these benefits are both complementary and cumulative in overall effect on farm health. Effective organic pest control requires a thorough understanding of pest life cycles and interactions.

Organic pest control is similar to integrated pest management in some respects.

Livestock

Raising livestock and poultry, for meat, dairy and eggs, is another traditional, farming activity that complements growing. Organic farms attempt to provide animals with "natural" living conditions and feed. While the USDA does not require any animal welfare requirements be met for a product to be marked as organic, this is a variance from older organic farming practices. Ample, free-ranging outdoor access, for grazing and exercise, is a distinctive feature, and crowding is avoided. Feed is also organically grown, and drugs, including antibiotics, are not ordinarily used (and are prohibited under organic regulatory regimes). Animal health and food quality are thus pursued in a holistic "fresh air, exercise, and good food" approach.

Also, horses and cattle used to be a basic farm feature that provided labour, for hauling and plowing, fertility, through recycling of manure, and fuel, in the form of food for farmers and other animals. While today, small growing operations often do not include livestock, domesticated animals are a desirable part of the organic farming equation, especially for true sustainability, the ability of a farm to function as a self-renewing unit.

Organic Farming Systems

There are several organic farming systems. Biodynamic farming is a comprehensive approach, with its own international governing body. The Do Nothing Farming method focuses on a minimum of mechanical cultivation and labour for grain crops. French intensive and biointensive, methods are well-suited to organic principles. Other techniques are permaculture and no-till farming. Finally, newcomers as the Agro-ecologic system focus on a blend of a more large-scale approach with imbedded natural/organic farming techniques. A farm may choose to adopt a particular method, or a mix of techniques.

While fundamentally different, large-scale agriculture and organic farming are not entirely mutually exclusive. For example, Integrated

Pest Management is a multifaceted strategy that can include synthetic pesticides as a last resort—both organic and conventional farms use IPM systems for pest control.

Soil Management

Plants need nitrogen, phosphorus, and potassium, as well as micronutrients and symbiotic relationships with fungi and other organisms to flourish, but getting enough nitrogen, and particularly synchronisation so that plants get enough nitrogen at the right time (when plants need it most), is likely the greatest challenge for organic farmers. Crop rotation and green manure ("cover crops") help to provide nitrogen through legumes (more precisely, the *Fabaceae* family) which fix nitrogen from the atmosphere through symbiosis with rhizobial bacteria. Intercropping, which is sometimes used for insect and disease control, can also increase soil nutrients, but the competition between the legume and the crop can be problematic and wider spacing between crop rows is required. Crop residues can be ploughed back into the soil, and different plants leave different amounts of nitrogen, potentially aiding synchronisation. Organic farmers also use animal manure, certain processed fertilizers such as seed meal and various mineral powders such as rock phosphate and greensand, a naturally occurring form of potash which provides potassium. Together these methods help to control erosion. In some cases pH may need to be amended. Natural pH amendments include lime and sulfur, but in the U.S. some compounds such as iron sulfate, aluminium sulfate, magnesium sulfate, and soluble boron products are allowed in organic farming.

Mixed farms with both livestock and crops can operate as ley farms, whereby the land gathers fertility through growing nitrogen-fixing forage grasses such as white clover or alfalfa and grows cash crops or cereals when fertility is established. Farms without livestock ("stockless") may find it more difficult to maintain fertility, and may rely more on external inputs such as imported manure as well as grain legumes and green manures, although grain legumes may fix limited nitrogen because they are harvested. Horticultural farms growing fruits and vegetables which operate in protected conditions are often even more reliant upon external inputs.

Biological research on soil and soil organisms has proven beneficial to organic farming. Varieties of bacteria and fungi break down chemicals, plant matter and animal waste into productive soil nutrients. In turn, they produce benefits of healthier yields and more productive soil for future crops. Fields with less or no manure display significantly lower yields, due to decreased soil microbe community, providing a healthier, more arable soil system.

Weed Management

Organic weed management promotes weed suppression, rather than weed elimination, by enhancing crop competition and phytotoxic effects on weeds. Organic farmers integrate cultural, biological, mechanical, physical and chemical tactics to manage weeds without synthetic herbicides.

Organic standards require rotation of annual crops, meaning that a single crop cannot be grown in the same location without a different, intervening crop. Organic crop rotations frequently include weed-suppressive cover crops and crops with dissimilar life cycles to discourage weeds associated with a particular crop. Organic farmers strive to increase soil organic matter content, which can support microorganisms that destroy common weed seeds.

Other cultural practices used to enhance crop competitiveness and reduce weed pressure include selection of competitive crop varieties, high-density planting, tight row spacing, and late planting into warm soil to encourage rapid crop germination.

Mechanical and physical weed control practices used on organic farms can be broadly grouped as:

- Tillage - Turning the soil between crops to incorporate crop residues and soil amendments; remove existing weed growth and prepare a seedbed for planting;
- Cultivation - Disturbing the soil after seeding;
- Mowing and cutting - Removing top growth of weeds;
- Flame weeding and thermal weeding - Using heat to kill weeds; and
- Mulching - Blocking weed emergence with organic materials, plastic films, or landscape fabric.

Some naturally sourced chemicals are allowed for herbicidal use. These include certain formulations of acetic acid (concentrated vinegar), corn gluten meal, and essential oils. A few selective bioherbicides based on fungal pathogens have also been developed. At this time, however, organic herbicides and bioherbicides play a minor role in the organic weed control toolbox.

Weeds can be controlled by grazing. For example, geese have been used successfully to weed a range of organic crops including cotton, strawberries, tobacco, and corn, reviving the practice of keeping cotton patch geese, common in the southern U.S. before the 1950s. Similarly, some rice farmers introduce ducks and fish to wet paddy fields to eat both weeds and insects.

Advance Sowing

Advance Sowing (also known as "no kill cropping") is an agricultural method developed by Bruce Maynard in 1996 in NSW, Australia that allows the production of annual crops from perennial grasslands. It consists in dry-sowing crops directly into existing pastures without using tillage, fertilizer or chemicals.

Principles

Advance Sowing has 5 major principles:

1. Sowing is done when the topsoil is dry.
2. Coulter type sowing equipment must be used.
3. No Herbicidesor pesticides are applied at any stage.
4. No Fertilizers are applied at any stage.
5. Grazing management must be good.

The rationale behind the method is to produce crops without simplifying the biodiversity. All other commonly used sowing methods of cropping rely on eliminating some or all of the plant and animals present to create an advantage for the growing crop. Advance Sowing relies on complementarity of plant/animal interactions to produce biomass that can be utilised directly for human consumption or fed to animals.

No Kill Cropping offers opportunities to expand the grain crop production throughout the world as an addition to the grazing resources that people depend upon. By using No Kill in areas that are already grasslands or are arid or highly erodible grain can be produced without risking ecological damage. Other advantages include the low capital costs involved and also low needs for outside energy intensive inputs.

No Kill Cropping is mentioned in the book "Here On Earth" by Tim Flannery, 2010. It is described as 'Zero Kill' or 'Zero Till' on pages 264 and 268. Flannery confuses the method by including the use of ploughs and also of the term pastures rather than grasslands that the crop is sown into. He proposes that the system has possibilities to promote "coevolution's capacity to increase biological productivity and ecosystem stability".

Controlling Other Organisms

Organisms aside from weeds that cause problems on organic farms include arthropods (e.g., insects, mites), nematodes, fungi and bacteria. Organic farmers use a wide range of Integrated Pest Management practices to prevent pests and diseases. These include, but are not limited to, crop rotation and nutrient management; sanitation to remove pest habitat; provision of habitat for beneficial organisms; selection of pest-resistant crops and animals; crop protection using physical barriers, such as row covers; and crop diversification through companion planting or establishment of polycultures.

Organic farmers often depend on biological pest control, the use of beneficial organisms to reduce pest populations. Examples of beneficial insects include minute pirate bugs, big-eyed bugs, and to a lesser extent ladybugs (which tend to fly away), all of which eat a wide range of pests. Lacewings are also effective, but tend to fly away. Praying mantis tend to move more slowly and eat less heavily. Parasitoid wasps tend to be effective for their selected prey, but like all small insects can be less effective outdoors because the wind controls their movement. Predatory mites are effective for controlling other mites.

When these practices are insufficient to prevent or control pests an organic farmer may apply a pesticide. With some exceptions, naturally occurring pesticides are allowed for use on organic farms, and synthetic substances are prohibited. Pesticides with different modes of action should be rotated to minimise development of pesticide resistance.

Naturally derived insecticides allowed for use on organic farms use include *Bacillus thuringiensis* (a bacterial toxin), pyrethrum (a chrysanthemum extract), spinosad (a bacterial metabolite), neem (a tree extract) and rotenone (a legume root extract). These are sometimes called green pesticides because they are generally, but not necessarily, safer and more environmentally friendly than synthetic pesticides. Rotenone and pyrethrum are particularly controversial because they work by attacking the nervous system, like most conventional insecticides. Fewer than 10% of organic farmers use these pesticides regularly; one survey found that only 5.3% of vegetable growers in California use rotenone while 1.7% use pyrethrum (Lotter 2003:26).

Naturally derived fungicides allowed for use on organic farms include the bacteria *Bacillus subtilis* and *Bacillus pumilus*; and the fungus *Trichoderma harzianum*. These are mainly effective for diseases affecting roots. Agricultural Research Service scientists have found that caprylic acid, a naturally occurring fatty acid in milk and coconuts, as well as other natural plant extracts have antimicrobial characteristics that can help. Compost tea contains a mix of beneficial microbes, which may attack or out-compete certain plant pathogens, but variability among formulations and preparation methods may contribute to inconsistent results or even dangerous growth of toxic microbes in compost teas.

Some naturally derived pesticides are not allowed for use on organic farms. These include nicotine sulfate, arsenic, and strychnine.

Synthetic pesticides allowed for use on organic farms include insecticidal soaps and horticultural oils for insect management; and Bordeaux mixture, copper hydroxide and sodium bicarbonate for managing fungi.

Biological Pest Control

Biological control of pests in agriculture is a method of controlling pests (including insects, mites, weeds and plant diseases) that relies on predation, parasitism, herbivory, or other natural mechanisms. It can be an important component of integrated pest management (IPM) programs.

Biological control is defined as the reduction of pest populations by natural enemies and typically involves an active human role. Natural enemies of insect pests, also known as biological log control agents; and include predators, parasitoids, and pathogens. Biological control agents of plant diseases are most often referred to as antagonists. Biological control agents of weeds include herbivores and plant pathogens. Predators, such as birds, lady beetles and lacewings, are mainly free-living species that consume a large number of prey during their whole lifetime. Parasitoids are species whose immature develops on or within a single insect host, ultimately killing or fatally infecting the host. Most have a very narrow host range. Many species of wasps and some flies are parasitoids. Pathogens are disease-causing organisms including bacteria, fungi, and viruses. They kill or debilitate their own host and are relatively specific. There are three basic types of biological control strategies; conservation, classical biological control, and augmentation.

Conservation

The conservation of natural enemies is probably the most important and readily available biological control practice available to homeowners and gardeners. Natural enemies occur in all areas, from the private garden to the open field. They are adapted to the habitat and to the target pest, and their conservation is generally simple and cost-effective. Lacewings, lady beetles, hover fly larvae, and parasitized aphid mummies are almost always present in aphid colonies. Fungus-infected adult flies are often common following periods of high humidity. These naturally occurring biological controls are often susceptible to the same pesticides used to target their pest hosts. Preventing the accidental eradication of natural enemies is termed "simple conservation."

To conserve and encourage pest insect eating birds, including native plants and ornamental plants that supply berries, acorns, nuts, seeds, nectar, and other vegetative foods, and also bird nest building materials, encourages their presence, health, and new generations. These qualities can also increase the visible population to enjoy in a garden. Using companion planting and the birds' insect cuisine habits is a traditional method for biological control agent pest control in an organic garden and any landscape, and in organic farming and sustainable agriculture. Installing specified nest boxes for mosquito-eating bats reduces a pest and increases endangered species conservation.

Effects of Biological Control

Effects on Native Biodiversity: Biological control can potentially have positive and negative effects on biodiversity. Usually a biological control is introduced to an area to protect a native species from an invasive or exotic species that has moved into its area. The control is introduced to lessen the competition between native and introduced species. However, the introduced control does not always target only the intended species. It can also target native species.

When introducing a high biological control to a new area, a primary concern is the host- or prey-specificity of the control agent. Generalist feeders (control agents that are not restricted to a single species or a small range of species) often make poor biological control agents, and may become invasive species themselves. For this reason, potential biological control agents should be subject to extensive testing and quarantine before release into any new environment. If a species is introduced and attacks a native species, the biodiversity in that area can change dramatically. When one native species is removed from an area, it may have filled an essential ecological niche. When this niche is absent it may directly affect the entire ecosystem.

Because they tend to be generalist feeders, vertebrate animals seldom make good biological control agents, and many of the classic cases of "biocontrol gone awry" involve vertebrates. For example, the cane toad, *Bufo marinus*, was introduced as a biological control and had significant negative impact on biodiversity. The cane toad was intentionally introduced to Australia to control the introduced French's Cane Beetle and the Greyback Cane Beetle. When introduced, the cane toad thrived very well and did not only feed on cane beetles but other insects as well. The cane toad soon spread very rapidly, thus taking over native amphibian habitat. The introduction of the cane toad also brought foreign disease to native reptiles. This drastically reduced the population of native toads and frogs. "The cane toad also exudes and can squirt poison from the parotid glands on their shoulders when threatened or handled. This toxin contains a cocktail of chemicals that can kill animals that eat it. Freshwater crocodiles, goannas, tiger snakes, dingos and northern quolls have all died after eating cane toads, as have pet dogs (Cane toad, 2003). This example shows how a small mis-introduced organism can alter the native biodiversity in a large ecosystem in a very expedient manner. A pyramid effect can take place if native species are reduced or eradicated. The domino effect keeps on going and can potentially exude on other bordering ecosystems until an equilibrium is reached.

A second example of a biological control agent that subsequently crossed over to native species is the *Rhinocyllus conicus*. The seed feeding weevil was introduced to North America to control exotic thistles (Musk and Canadian). However, the weevil did not target only the exotic thistles, it also targeted native thistles that are essential to various native insects. The native insects rely solely on native thistles and do not adapt to other plant species. Therefore, they cannot survive. Biological controls do not always have negative impacts on biodiversity (Corry 2000).

Successful biological control reduces the population density of the target species over several years, thus providing the potential for native species to re-establish. In addition, regeneration and re-establishment programs can aid to the recovery of native species. Native species can be affected in a positive way as well. To develop or find a biological control that exerts control only on the targeted species is a very lengthy process of research and experiments. In the late 19th century, the citrus industry was in great fear when the cottony cushion scale was discovered. This organism could cause a great deal of economic loss to the industry. However, a biological control was introduced.

The vedalia beetle and a parasitoid fly were introduced to control the pest. Within a few years time, the cottony cushion scale was controlled by the natural enemies and the citrus industry suffered little financial loss. Many exotic or invasive species can suppress the development of native species. The introduction of an effective biological control that reduces the population of the invasive species allows the rejuvenation of the native species. Biological controls can reduce competition for biotic and abiotic factors which can result in the re-establishment of the once over ran native species.

Effects on Invasive Species

Figure 1: The invasive species *Alternanthera philoxeroides* (alligator weed) was controlled in Florida (U.S.) by the introduction of *Agasicles hygrophila* (alligator weed flea beetle)

Invasive species are closely associated with biological controls because the environment in which they are invasive most likely does not contain their natural enemies. If invasive species are not controlled, biodiversity may be at great threat in the affected area. An example of an invasive species is the alligator weed. This plant was introduced to the United States from South America. This aquatic weed spreads rapidly and causes many problems in lakes and rivers. The weed takes root in shallow water causing major problems for navigation, irrigation, and flood control. The alligator weed flea beetle and two other biological controls were released in Florida. Because of their success, Florida banned the use of herbicides to control alligator weed three years after the controls were introduced. Similarly, *Galerucella calmariensis*, a leaf beetle, has been introduced in North America as a control agent for purple loosestrife (*Lythrum salicaria*).

Biological controls for invasive species also can have a negative impact on biodiversity. The cane toad, as mentioned previously, is an example of trying to control an invasive species. The cane toad was introduced to eradicate an invasive species. It became invasive, thus altering the biodiversity. The introduction of the cane toad could have caused more of a disturbance in biodiversity than the targeted species did.

Classical Biological Control

Classical biological control is the introduction of natural enemies to a new locale where they did not originate or do not occur naturally. This is usually done by government authorities. In many instances the complex of natural enemies associated with an insect pest may be inadequate. This is especially evident when an insect pest is accidentally introduced into a new geographic area without its associated natural enemies. These introduced pests are referred to as exotic pests and comprise about 40% of the insect pests in the United States. Examples of introduced vegetable pests include the European corn borer (*Ostrinia nubilalis*), one of the most destructive insects in North America. To obtain the needed natural enemies, scientists turned to classical biological control. This is the practice of importing, and releasing for establishment, natural enemies to control an introduced (exotic) pest, although it is also practiced against native insect pests. The first step in the process is to determine the origin of the introduced pest and then collect appropriate natural enemies associated with the pest or closely related species. The natural enemy is then passed through a rigorous quarantine process, to ensure that no unwanted organisms (such as hyperparasitoids) are introduced, then they are mass produced, and released. Follow-up studies are conducted to determine if the natural enemy becomes successfully established at the site of release, and to assess the long-term benefit of its presence.

There are many examples of successful classical biological control programs.

Joseph Needham noted a Chinese text dating from 304AD, *Records of the Plants and Trees of the Southern Regions*, by Hsi Han, which describes mandarin oranges protected by biological pest control techniques that are still in use today.

One of the earliest successes in the west was in controlling *Icerya purchasi*, the cottony cushion scale, a pest that was devastating the California citrus industry in the late 19th century. A predatory insect *Rodolia cardinalis* (the Vedalia Beetle), and a parasitoid fly were introduced from Australia by Charles Valentine Riley. Within a few years the cottony cushion scale was completely controlled by these introduced natural enemies.

Damage from *Hypera postica* Gyllenhal, the alfalfa weevil, a serious introduced pest of forage, was substantially reduced by the introduction of several natural enemies. 20 years after their introduction the population of weevils in the alfalfa area treated for alfalfa weevil in the Northeastern United States was reduced by 75 percent. A small wasp, *Trichogramma ostriniae*, was introduced from China to help control the European corn borer making it a recent example of a long history of classical biological control efforts for this major pest. Many classical biological control programs for insect pests and weeds are under way across the United States and Canada. The population of *Levuana irridescens* (the Levuana moth), a serious coconut pest in Fiji, was brought under control by a classical biological control program in the 1920s.

Classical biological control is long lasting and inexpensive. Other than the initial costs of collection, importation, and rearing, little expense is incurred. When a natural enemy is successfully established it rarely requires additional input and it continues to kill the pest with no direct help from humans and at no cost. Classical biological control does not always work. It is usually most effective against exotic pests and less so against native insect pests. The reasons for failure are not often known but may include the release of too few individuals, poor adaptation of the natural enemy to environmental conditions at the release location, and lack of synchrony between the life cycle of the natural enemy and host pest.

Augmentation

This third type of biological control involves the supplemental release of natural enemies. Relatively few natural enemies may be released at a critical time of the season (inoculative release) or literally millions may be released (inundative release). Additionally, the cropping system may be modified to favour or augment the natural enemies. This latter practice is frequently referred to as habitat manipulation. An example of inoculative release occurs in greenhouse production of several crops. Periodic releases of the parasitoid, *Encarsia formosa*, are used to control greenhouse whitefly, and the predaceous mite, *Phytoseiulus persimilis*, is used for control of the two-spotted spider mite.

Lady beetles, lacewings, or parasitoids such as those from the genus *Trichogramma* are frequently released in large numbers (inundative release). Recommended release rates for Trichogramma in vegetable or field crops range from 5,000 to 200,000 per acre (1 to 50 per square metre) per week depending on level of pest infestation. Similarly, entomopathogenic nematodes are released at rates of millions and even billions per acre for control of certain soil-dwelling insect pests.

The spraying of octopamine analogues (such as 3-FMC) has been suggested as a way to boost the effectiveness of augmentation. Octopamine, regarded as the invertebrate counterpart of dopamine plays a role in activating the insects' flight-or-fight response. The idea behind using octopamine analogues to augment biological control is that natural enemies will be more effective in their eradication of the pest, since the pest will be behaving in an unnatural way because its flight-or-fight mechanism has been activated. Octopamine analogues are purported to have two desirable characteristics for this type of application: (1) they affect insects at very low dosages (2) they do not have a physiological effect in humans (or other vertebrates).

Habitat or environmental manipulation is another form of augmentation. This tactic involves altering the cropping system to augment or enhance the effectiveness of a natural enemy. Many adult parasitoids and predators benefit from sources of nectar and the protection provided by refuges such as hedgerows, cover crops, and weedy borders. Also, the provisioning of natural shelters in the form of wooden caskets, boxes or (turnaround) flowerpots is a form of this. For example, the stimulation of the natural predator *Dermaptera* is done in gardens by hanging up turnaround flowerpots with straw or wood wool.

Mixed plantings and the provision of flowering borders can increase the diversity of habitats and provide shelter and alternative food sources. They are easily incorporated into home gardens and even small-scale commercial plantings, but are more difficult to accommodate in large-scale crop production. There may also be some conflict with pest control for the large producer because of the difficulty of targeting the pest species and the use of refuges by the pest insects as well as natural enemies.

Examples of habitat manipulation include growing flowering plants (pollen and nectar sources) such as Buckwheat near crops to attract and maintain populations of natural enemies. For example, hover fly adults can be attracted to umbelliferous plants in bloom.

Biological control experts in California have demonstrated that planting prune trees in grape vineyards provides an improved overwintering habitat or refuge for a key grape pest parasitoid. The prune trees harbour an alternate host for the parasitoid, which could previously overwinter only at great distances from most vineyards. Caution should be used with this tactic because some plants attractive to natural enemies may also be hosts for certain plant diseases, especially plant viruses that could be vectored by insect pests to the crop. Although the tactic appears to hold much promise, only a few examples have been adequately researched and developed.

Examples of Predators

Figure 2: Lacewings are available from biocontrol dealers.

Ladybugs, and in particular their larvae which are active between May and July in the northern hemisphere, are voracious predators of aphids such as greenfly and blackfly, and will also consume mites, scale insects and small caterpillars. The ladybug is a very familiar beetle with various coloured markings, whilst its larvae are initially small and spidery, growing up to 17 mm long. The larvae have a tapering segmented grey/black body with orange/yellow markings and ferocious mouthparts. They can be encouraged by cultivating a patch of nettles in the garden and by leaving hollow stems and some plant debris over winter so that they can hibernate.

Hoverflies resemble slightly darker bees or wasps and they have characteristic hovering, darting flight patterns. There are over 100 species of hoverfly whose larvae principally feed upon greenfly, one larva devouring up to fifty a day, or 1000 in its lifetime. They also eat fruit tree spider mites and small caterpillars. Adults feed on nectar and pollen, which they require for egg production. Eggs are minute (1 mm), pale yellow white and laid singly near greenfly colonies. Larvae are 8–17 mm long, disguised to resemble bird droppings, they are legless and have no distinct head. Semi-transparent in a range of colours from green, white, brown and black.

Figure 3: Predatory Polistes wasp looking for bollworms or other caterpillars on a cotton plant

Hoverflies can be encouraged by growing attractant flowers such as the poached egg plant *(Limnanthes douglasii)*, marigolds or phacelia throughout the growing season.

Dragonflies are important predators of mosquitoes, both in the water, where the dragonfly naiads eat mosquito larvae, and in the air, where adult dragonflies capture and eat adult mosquitoes. Community-wide mosquito control programs that spray adult mosquitoes also kill dragonflies, thus removing an important biocontrol agent, and can actually increase mosquito populations in the long term.

Other useful garden predators include lacewings, pirate bugs, rove and ground beetles, aphid midge, centipedes, spiders, predatory mites, as well as larger fauna such as frogs, toads, lizards, hedgehogs, slow-worms and birds. Cats and rat terriers kill field mice, rats, June bugs, and birds. Dogs chase away many types of pest animals. Dachshunds are bred specifically to fit inside tunnels underground to kill badgers.

More examples:

- *Phytoseiulus persimilis* (against spider mites)
- *Amblyseius californicus* (against spider mites)
- *Amblyseius cucumeris* (against spider mites)
- *Typhlodromips swirskii* (against spider mites, thrips, and white flies)
- *Feltiella acarisuga* (against spider mites)
- *Stethorus punctillum* (against spider mites)
- *Macrolophus caluginosus* (against spider mites).

Parasitoid Insects

Most insect parasitoids are wasps or flies. Parasitoids comprise a diverse range of insects that lay their egg on or in the body of an insect host, which is then used as a food for developing larvae. Parasitic wasps take much longer than predators to consume their victims, for if the larvae were to eat too fast they would run out of food before they became adults. Such parasites are very useful in the organic garden, for they are very efficient hunters, always at work searching for pest invaders. As adults they require high energy fuel as they fly from place to place, and feed upon nectar, pollen and sap, thereby pollinating plenty of flowering plants, particularly buckwheat, umbellifers, and composites will encourage their presence.

Four of the most important groups are:

- Ichneumonid wasps: (5–10 mm). Prey mainly on caterpillars of butterflies and moths.
- Braconid wasps: Tiny wasps (up to 5 mm) attack caterpillars and a wide range of other insects including greenfly. A common parasite of the cabbage white caterpillar- seen as clusters of sulphur yellow cocoons bursting from collapsed caterpillar skin.
- Chalcid wasps: Among the smallest of insects (<3 mm). Parasitize eggs/larvae of greenfly, whitefly, cabbage caterpillars, scale insects and Strawberry Tortrix Moth (*Acleris comariana*).

- Tachinid flies: Parasitize a wide range of insects including caterpillars, adult and larval beetles, true bugs, and others.

Examples of parasitoids: wasp

- *Encarsia formosa* (against white flies)
- *Eretmocerus* spp. (against white flies)
- *Aphidius colemani* (against aphids).

Biological Control with Micro-organisms

Various microbial insect diseases occur naturally, but may also be used as biological pesticides. When naturally occurring, these outbreaks are density-dependent in that they generally only occur as insect populations become denser.

Bacteria and Biological Control

Bacteria used for biological control infect insects via their digestive tracts, so insects with sucking mouth parts like aphids and scale insects are difficult to control with bacterial biological control. *Bacillus thuringiensis* is the most widely applied species of bacteria used for biological control, with at least four subspecies used to control Lepidopteran (moth, butterfly), Coleopteran (beetle) and Dipteran (true flies) insect pests.

Fungi and Biological Control

Fungi that cause disease in insects are known as entomopathogenic fungi, including at least fourteen species of entomophthoraceous fungi attack aphids. Species in the genus *Trichoderma* are used to manage some soilborne plant pathogens. *Beauveria bassiana* is used to manage different types of pest such whiteflies, thrips, aphids and weevils.

Examples of entomopathogenic fungi:

- *Beauveria bassiana* (against white flies, thrips, aphids and weevils)
- *Paecilomyces fumosoroseus* (against white flies, thrips and aphids)
- *Metarhizium* spp. (against beetles, locusts, Hemiptera, spider mites and other pests)
- *Lecanicillium lecanii* (against white flies, thrips and aphids)
- *Cordyceps* species (sometines teleomorphs of the above: that infect a wide spectrum of arthropods).

Combined use of Parasitoids and Pathogens

In cases of massive and severe infection of invasive pests, techniques of pest control are often used in combination. An example being, that of the emerald ash borer (*Agrilus planipennis* Fairmaire, family Buprestidae), an invasive beetle from China, which has destroyed tens of millions of ash trees in its introduced range in North America's part of the campaign against the emerald ash borer (EAB), American scientists in conjunction with the Chinese Academy of Forestry searched since 2003 for its natural enemies in the wild leading to the discovery of several parasitoid wasps, namely *Tetrastichus planipennisi*, a gregarious larval endoparasitoid, *Oobius agrili*, a solitary, parthenogenic egg parasitoid, and *Spathius agrili*, a gregarious larval ectoparasitoid. These have been introduced and released into the United States of America as a possible biological control of the emerald ash borer. Initial results have shown promise with *Tetrastichus planipennisi* and it is now being released along with *Beauveria bassiana*, a fungal pathogen with known insecticidal properties.

Plants to Regulate Insect Pests

Choosing a diverse range of plants for the garden can help to regulate pests in a variety of ways, including;

- Masking the crop plants from pests, depending on the proximity of the companion or intercrop.
- Producing olfactory inhibitors, odours that confuse and deter pests.
- Acting as trap plants by providing an alluring food that entices pests away from crops.
- Serving as nursery plants, providing breeding grounds for beneficial insects.
- Providing an alternative habitat, usually in a form of a shelterbelt, hedgerow, or beetle bank where beneficial insects can live and reproduce. Nectar-rich plants that bloom for long periods are especially good, as many beneficials are nectivorous during the adult stage, but parasitic or predatory as larvae. A good example of this is the soldier beetle which is frequently found on flowers as an adult, but whose larvae eat aphids, caterpillars, grasshopper eggs, and other beetles.
- Some plants have chemical defences in order to regulate pests. The geranium has developed such a defence against Japanese

beetles, one of the most damaging and expensive pests to control when it comes to ornamental and turf plants. The geranium's petals contain a chemical compound that paralyzes the beetle within 30 minutes of ingestion. The beetle will remain paralyzed for several hours and will typically regain movement within 24 hours. However, while paralyzed the beetle is very vulnerable to its predators and is usually hunted before the paralysis subsides. Agricultural Research Service (ARS) scientists are working to isolate the chemical compound in geraniums that causes the paralysis in the beetles. Scientists hope to one day use this natural pesticide to control the population of beetles. In addition to this research ARS scientists are studying ways to help geranium leaves better hold on to pesticide chemicals that are sprayed on them, that way less pesticides will have to be applied to the leaves.

Plants to Regulate Plants

The legume vine *Mucuna pruriens* is used in the countries of Benin and Vietnam as a biological control for problematic *Imperata cylindrica* grass. *Mucuna pruriens* is said not to be invasive outside its cultivated area. *Desmodium uncinatum* can be used in push-pull farming to stop the parasitic plant, *Striga*.

Directly Introducing Biological Controls

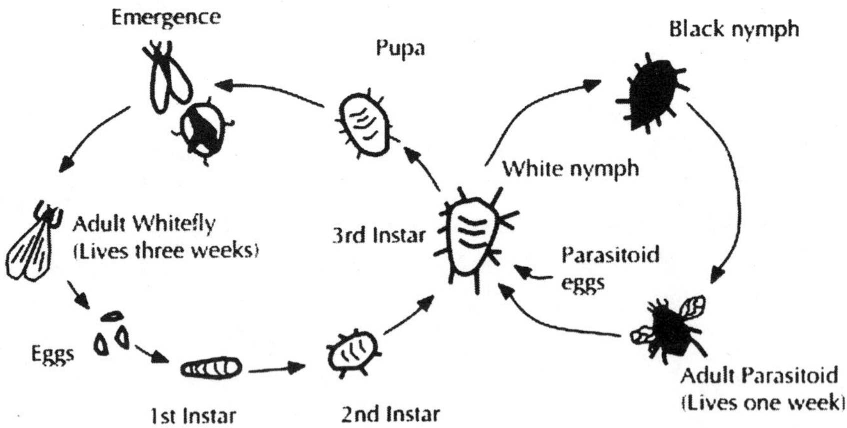

Figure 4: Diagram illustrating the life cycles of Greenhouse whitefly and its parasitoid wasp Encarsia formosa

Most of the biological controls listed above depend on providing incentives in order to 'naturally' attract beneficial insects to the garden.

However there are occasions when biological controls can be directly introduced. Common biocontrol agents include parasitoids, predators, pathogens or weed feeders. This is particularly appropriate in situations such as the greenhouse, a largely artificial environment, and are usually purchased by mail order.

Some biocontrol agents that can be introduced include;

- *Encarsia formosa.* This is a small predatory chalcid wasp which is parasitical on whitefly, a sap-feeding insect which can cause wilting and black sooty moulds. It is most effective when dealing with low level infestations, giving protection over a long period of time. The wasp lays its eggs in young whitefly 'scales', turning them black as the parasite larvae pupates. It should be introduced as soon as possible after the first adult whitefly are seen. Should be used in conjunction with insecticidal soap.
- Red spider mite, another pest found in the greenhouse, can be controlled with the predatory mite *Phytoseilus persimilis.* This is slightly larger than its prey and has an orange body. It develops from egg to adult twice as fast as the red spider mite and once established quickly overcomes infestation.
- A fairly recent development in the control of slugs is the introduction of 'Nemaslug', a microscopic nematode (*Phasmarhabditis hermaphrodita*) which will seek out and parasitize slugs, reproducing inside them and killing them. The nematode is applied by watering onto moist soil, and gives protection for up to six weeks in optimum conditions, Nemaslug nematodes are mainly effective with small and young slugs under the soil surface.
- A bacterial biological control which can be introduced in order to control butterfly caterpillars is *Bacillus thuringiensis.* This available in sachets of dried spores which are mixed with water and sprayed onto vulnerable plants such as brassicas and fruit trees. The bacterial disease will kill the caterpillars, but leave other insects unharmed. There are strains of *Bt* that are effective against other insect larvae. *Bt israelensis* is effective against mosquito larvae and some midges.
- A viral biological control which can be introduced in order to control the overpopulation of European rabbit in Australia is

the rabbit haemorrhagic disease virus that causes the rabbit haemorrhagic disease.

- A biological control being developed for use in the treatment of plant disease is the fungus *Trichoderma viride.* This has been used against Dutch Elm disease, and to treat the spread of fungal and bacterial growth on tree wounds. It may also have potential as a means of combating silver leaf disease.
- Several species of dung beetle were introduced to Australia from South Africa and Europe during the Australian Dung Beetle Project (1965–1985) led by Dr. George Bornemissza of the Commonwealth Scientific and Industrial Research Organisation in order to biologically control the population of pestilent bush flies and parasitic worms.
- The parasitoid *Gonatocerus ashmeadi* (Hymenoptera: Mymaridae) has been introduced to control the glassy-winged sharpshooter *Homalodisca vitripennis* (Hemipterae: Cicadellidae) in French Polynesia and has successfully controlled ~95% of the pest density.

Negative Consequences of Biological Pest Control

In some cases, biological pest control could have unforeseen averse consequences that outweigh all benefits, often by becoming an invasive species. For example:

- When the mongoose was introduced to Hawaii in order to control the rat population, it preyed on the endemic birds of Hawaii, especially their eggs, more often than it ate the rats. (Note, however, that the introduction of the mongoose was not undertaken based on scientific—or perhaps any—understanding of the consequences of such an action. The introduction of a generalist mammal for biocontrol of anything would be unthinkable by any reasonable standards today.)
- Cane toads (*Bufo marinus*) were introduced to Australia in the 1930s in a failed attempt to control the cane beetle, a pest of sugar cane crops. 102 toads were obtained from Hawaii and bred in captivity to increase their numbers until they were released into the sugar cane fields of the tropic north in 1935. It was later discovered that the toads could not jump very high and so they could not eat the cane beetles which stayed up on the upper stalks of the cane plants. The toad population grew dramatically and eventually out-competed native species. Not only were these toads very harmful to the Australian

environment, they were also very toxic to would-be predators such as the native snakes.

- 5 cats brought to the subantarctic Marion Islands to catch mice in 1949 multiplied to 3,400 in about two decades and started to threaten local extinction of birds. They had to be infected with feline distemper virus. The rest were shot and completely eliminated by the 1990s.
- The sturdy and prolific mosquito fish was introduced from around the Gulf of Mexico to around the world in the 1930s and 40s to combat malaria; however, it was found to cause the decline of local fish and frogs through competition of other food source as well as eating their eggs.

Living organisms, through the process of evolution, may achieve increased resistance to biological, chemical, and physical methods of control over time. In the event the target pest population is not completely exterminated or is still capable of reproduction (were the pest control means a form of sterilization), the surviving population could acquire a tolerance to the applied pressures - this can result in an evolutionary arms race with the control method.

Beneficial Insects

Beneficial insects (sometimes called beneficial bugs) are any of a number of species of insects that perform valued services like pollination and pest control. The concept of *beneficial* is subjective and only arises in light of desired outcomes from a human perspective. In farming and agriculture, where the goal is to raise selected crops, insects that hinder the production process are classified as pests, while insects that assist production are considered beneficial. In horticulture and gardening; pest control, habitat integration, and 'natural vitality' aesthetics are the desired outcome with beneficial insects.

Encouraging beneficial insects, by providing suitable living conditions, is a pest control strategy, often used in organic farming, organic gardening or Integrated Pest Management. Companies specialising in biological pest control sell many types of beneficial insects, particularly for use in enclosed areas, like greenhouses.

Types of Beneficial Insects

Some species of bee are beneficial as pollinators, although generally only efficient at pollinating plants from the same area of origin, facilitating propagation and fruit production for many plants. Also, some bees are predatory or parasitic, killing pest insects. This group includes not only honeybees, but also many other kinds that are more efficient at pollinating. Bees can be attracted by many companion plants, especially bee balm and pineapple sage for honeybees, or a piaceae like queen anne's lace and parsley, for predatory bees.

Figure 5: A European mantis hunts for prey on a shrub rose.

Ladybugs are generally thought of as beneficial because they eat large quantities of aphids, mites and other arthropods that feed on various plants.

Other insects commonly identified as beneficial include:

- Minute pirate bug
- Big eyed-bug
- Assassin bug
- Damsel bug
- Mealybug destroyer
- Soldier beetle
- Green lacewing
- Syrphid fly
- Tachinid fly
- Ichneumon wasp
- Trichogramma wasp.

Attractive Plants

- Plants in the Apiaceae and Asteraceae families are general valuable companions.

Companion Planting

Companion planting is the planting of different crops in proximity (in gardening and agriculture), on the theory that they assist each other in nutrient uptake, pest control, pollination, and other factors necessary to increasing crop productivity. Companion planting is a form of polyculture.

Companion planting is used by farmers and gardeners in both industrialised and developing countries for many reasons. Many of the modern principles of companion planting were present many centuries ago in cottage gardens in England and home gardens in Asia.

For farmers using an integrated pest management system, increased yield and/or reduction of pesticides is the goal. For gardeners, the combinations of plants also make for a more varied, attractive vegetable garden, as well as allowing more productive use of space.

History

In China, the mosquito fern has been used for at least one thousand years, as a companion plant for rice crops. It hosts a special cyanobacteria that fixes nitrogen from the atmosphere, and also blocks out light from getting to any competing plants, aside from the rice, which is planted when tall enough to stick out of the water above the azolla layer.

Companion planting was practiced in various forms by Native Americans prior to the arrival of Europeans. One common system was the planting of corn (maize) and pole beans together. The cornstalk would serve as a trellis for the beans to climb while the beans would fix nitrogen for the corn. The inclusion of squash with these two plants completes the Three Sisters technique, pioneered by Native American peoples.

Companion planting was widely touted in the 1970s as part of the organic gardening movement. It was encouraged for pragmatic reasons, such as natural trellising, but mainly with the idea that different species of plant may thrive more when close together. It is also a technique frequently used in permaculture, together with mulching, polyculture, and changing of crops.

Examples of Companion Plants

Nasturtium are well-known to attract caterpillars, therefore, planting them around vegetables such as lettuce or cabbage protects them from damage, as egg-laying insects will tend to prefer the nasturtium. This practice is called trap cropping.

Marigolds assist crops suffering from aphids (greenfly among others) through their smell deterring to aphids and attracting hoverflies (a predator of aphids).

Versions

There are a number of systems and ideas using companion planting.

Square foot gardening, for example, attempts to protect plants from many normal gardening problems by packing them as closely together as possible, which is facilitated by using companion plants, which can be closer together than normal.

Another system using companion planting is the forest garden, where companion plants are intermingled to create an actual ecosystem, emulating the interaction of up to seven levels of plants in a forest or woodland.

Organic gardening often depends on companion planting for its best performance, since so many synthetic means of fertilizing, weed reduction, pest control, and other garden needs are forbidden.

Good Weeds

There are many beneficial weeds, which can be allowed to grow alongside plants, imparting the same kinds of benefits as mixing cultivated crops.

Host-finding Disruption

Recent studies on host-plant finding have shown that flying pests are far less successful if their host-plants are surrounded by any other plant or even "decoy-plants" made of green plastic, cardboard, or any other green material.

The host-plant finding process occurs in phases:

- The first phase is stimulation by odours characteristic to the host-plant. This induces the insect to try to land on the plant it seeks. But insects avoid landing on brown (bare) soil. So if only the host-plant is present, the insects will quasi-systematically find it by simply landing on the only green thing around. This is called (from the point of view of the insect) "appropriate landing". When it does an "inappropriate landing", it flies off to any other nearby patch of green. It eventually leaves the area if there are too many 'inappropriate' landings.

- The second phase of host-plant finding is for the insect to make short flights from leaf to leaf to assess the plant's overall suitability. The number of leaf-to-leaf flights varies according to the insect species and to the host-plant stimulus received from each leaf. The insect must accumulate sufficient stimuli from the host-plant to lay eggs; so it must make a certain number of consecutive 'appropriate' landings. Hence if it makes an 'inappropriate landing', the assessment of that plant is negative, and the insect must start the process anew.

Thus it was shown that clover used as a ground cover had the same disruptive effect on eight pest species from four different insect orders. An experiment showed that 36% of cabbage root flies laid eggs beside cabbages growing in bare soil (which resulted in no crop), compared to only 7% beside cabbages growing in clover (which allowed a good crop). Simple decoys made of green cardboard also disrupted appropriate landings just as well as did the live ground cover.

Companion Plant Categories

Companion plants can benefit each other in a number of different ways, including:

- Flavour enhancement — some plants, especially herbs, seem to subtly change the flavour of other plants around them.
- Hedged investment — multiple plants in the same space increase the odds of some yield being given, even if one category encounters catastrophic issues
- Increased level interaction — plants that grow on different levels in the same space, perhaps providing ground cover or working as a trellis for another plant
- Nitrogen fixation — plants that fix nitrogen in the ground, making it available to other plants
- Pest suppression — repelling pest insects, weeds, nematodes, or pathogenic fungi, through chemical means .
- Pollinator and predator recruitment — The use of plants that produce copious nectar and protein-rich pollen in a vegetable garden (insectary plants) is a good way to recruit higher

populations of beneficial insects that control pests. Some insects in the adult form are nectar or pollen feeders, while in the larval form they are voracious predators of pest insects.

- Positive hosting — attracts or is inhabited by beneficial insects or other organisms which benefit plants, as with ladybugs or some "good nematodes".
- Protective shelter — one type of plant may serve as a wind break or shade for another
- Trap cropping — plants that attract pests away from others
- Pattern disruption — in a monoculture, pest spread from one plant to the next, is interrupted by companion plants.

Genetic Modification

A key characteristic of organic farming is the rejection of genetically engineered plants and animals. On October 19, 1998, participants at IFOAM's 12th Scientific Conference issued the Mar del Plata Declaration, where more than 600 delegates from over 60 countries voted unanimously to exclude the use of genetically modified organisms in food production and agriculture.

Although opposition to the use of any transgenic technologies in organic farming is strong, agricultural researchers Luis Herrera-Estrella and Ariel Alvarez-Morales continue to advocate integration of transgenic technologies into organic farming as the optimal means to sustainable agriculture, particularly in the developing world. Similarly, some organic farmers question the rationale behind the ban on the use of genetically engineered seed because they view this kind of biotechnology consistent with organic principles.

Although GMOs are excluded from organic farming, there is concern that the pollen from genetically modified crops is increasingly penetrating organic and heirloom seed stocks, making it difficult, if not impossible, to keep these genomes from entering the organic food supply. International trade restrictions limit the availability GMOs to certain countries.

The hazards that genetic modification could pose to the environment are hotly contested.

International Federation of Organic Agriculture Movements

The International Federation of Organic Agriculture Movements (IFOAM) is the worldwide umbrella organisation for the organic agriculture movement, uniting more than 750 member organisations in 108 countries. It declares its mission to be as follows:

"IFOAM's mission is leading, uniting and assisting the organic movement in its full diversity. Our goal is the worldwide adoption of ecologically, socially and economically sound systems that are based on the principles of Organic Agriculture."

Among its wide range of activities, IFOAM maintains an organic farming standard, and an organic accreditation service.

History

IFOAM began in Versailles, France, on November 5, 1972, during an international congress on organic agriculture organised by the French farmer organisation *Nature et Progrès*. The late Roland Chevriot, President of *Nature et Progrès*, took the initiative. There were 5 founding members representing different organisations: Lady Eve Balfour representing the Soil Association of Great Britain, Kjell Arman representing the Swedish Biodynamic Association, Pauline Raphaely representing the Soil Association of South Africa, Jerome Goldstein representing Rodale Press of the United States, and Roland Chevriot representing *Nature et Progrès* of France.

The aim of the new organisation was reflected in the name: *International Federation of Organic Agriculture Movements*. The founders hoped that the federation would meet what they saw as a major need: a unified, organised voice for organic food, and the diffusion and exchange of information on the principles and practices of organic agriculture across national and linguistic boundaries.

Structure

The IFOAM General Assembly serves as the foundation of IFOAM. It elects the IFOAM World Board for a three year term. The IFOAM World Board is a diverse group of individuals working voluntarily to lead IFOAM. The current World Board was elected at the IFOAM General Assembly in Vignola/Italy, which took place from June 22–24, 2008. The World Board appoints members to official committees, working groups and task forces based upon the recommendation of the IFOAM membership, and IFOAM member organisations also establish regional groups and sector specific interest groups.

International Standing

IFOAM actively participates in international agricultural and environmental negotiations with the United Nations and multilateral institutions to further the interests of the organic agricultural movement worldwide, and has observer status or is otherwise accredited by the following international institutions:

- ECOSOC Status with the United Nations General Assembly
- The Food and Agriculture Organisation of the United Nations (FAO)
- United Nations Conference on Trade and Development (UNCTAD)
- Codex Alimentarius Commission (FAO and World Health Organisation)
- United Nations Environment Program (UNEP)
- The Organisation for Economic Cooperation and Development (OECD)
- International Labour Organisation of the United Nations (ILO)
- International Organisation for Standardisation (ISO).

According to the One World Trust's Global Accountability Report 2008 which assessed a range of organisations in areas such as transparency, stakeholder participation and evaluation capacity, "IFOAM is the highest scoring international NGO, and at the top of the 30 organisations this year with a score of 71 percent".

Activities

IFOAM and GMOs: On October 19, 1998, participants at IFOAM's 12th Scientific Conference issued the Mar del Plata Declaration, where more than 600 delegates from over 60 countries voted unanimously to exclude the use of genetically modified organisms (GMOs) in food production and agriculture. From that point GMOs have been categorically excluded from organic farming.

Text of the Declaration

We, the undersigned participants at the 12th Scientific Conference of the International Federation of Organic Agriculture Movements (IFOAM) at Mar del Plata, Argentina, call on governments and regulatory agencies throughout the world to immediately ban the use of genetic engineering in agriculture and food production since it involves:

- Unacceptable threats to human health
- Negative and irreversible environmental impacts
- Release of organisms of an unrecallable nature

- Removal of the right of choice, both for farmers and consumers
- Violation of farmers' fundamental property rights and endangerment of their economic independence
- Practices, which are incompatible with the principles of sustainable agriculture as defined by IFOAM.

Signed by: Dr. Vandana Shiva (India), Hervé La Prairie (Outgoing IFOAM president, France), Linda Bullard (Incoming IFOAM president, USA/Belgium), Gunnar Rundgren (Incoming IFOAM vice-president, Sweden), Gerald Hermann (IFOAM Treasurer, Germany), Pipo Lernoud (Conference Coordinator, Argentina), Guillermo Schnitman (MAPO President, Argentina)

IFOAM and Training

Organic agriculture can contribute to meaningful socio-economic and ecologically sustainable development, especially in poorer countries. On one hand, this is due to the application of organic principles, which means efficient management of local resources (e.g., local seed varieties, manure, etc.) and therefore cost-effectiveness. On the other hand, the market for organic products – at local and international level – has tremendous growth prospects and offers creative producers and exporters in the South excellent opportunities to improve their income and living conditions. IFOAM is therefore active to give special support to the development of the Organic Agriculture Sector in Developing Countries through several means. Organic Agriculture is a very knowledge intensive production system. Therefore capacity building efforts play a central role in this regard. There are many efforts all around the world regarding the development of training material and the organisation of training courses related to Organic Agriculture. Existing knowledge is still scattered and not easy accessible. Especially in Developing Countries this situation remains an important constraint for the growth of the organic sector.

For that reason, IFOAM created an Internet training platform whose objective is to become the global reference point for Organic Agriculture training through free access to high quality training materials and training programs on Organic Agriculture.

In November 2007, the Training Platform hosted more than 170 free manuals and 75 training opportunities.

EU-Eco-regulation

The "European Union regulation (EEC) N° 2092/91 of the European Council of 24 June 1991 on organic production of agricultural products and indications referring thereto on agricultural products and foodstuffs" (EU-Eco-regulation) defines how agricultural products and foods that are designated as ecological products have to be grown. The regulation is derived from the guidelines of the International Federation of Organic Agriculture Movements (IFOAM), which is an association of about 750 member organisations in 108 countries.

In 1999 it was supplemented by regulation (EC) N° 1804/1999, which regulates the raising, labelling and inspection of the most relevant animal species (i.e. cattle, sheep, goats, horses and poultry). This agreement covers such issues as foodstuffs, disease prevention and veterinary treatments, animal welfare, husbandry practices and the management of manure. Genetically modified organisms (GMOs) and products derived from GMOs are explicitly excluded from organic production methods.

The regulation also regulates the import of ecologically grown agricultural products from third countries.

Organic Certification

Organic certification is a certification process for producers of organic food and other organic agricultural products. In general, any business directly involved in food production can be certified, including seed suppliers, farmers, [food] processors, retailers and restaurants.

Requirements vary from country to country, and generally involve a set of production standards for growing, storage, processing, packaging and shipping that include:

- avoidance of synthetic chemical inputs not on the National List of Allowed and Prohibited Substances(e.g. fertilizer, pesticides, antibiotics, food additives, etc.), genetically modified organisms, irradiation, and the use of biosolids;
- use of farmland that has been free from prohibited synthetic chemicals for a number of years (often, three or more);
- keeping detailed written production and sales records (audit trail);
- maintaining strict physical separation of organic products from non-certified products;
- undergoing periodic on-site inspections.

In some countries, certification is overseen by the government, and commercial use of the term *organic* is legally restricted. Certified organic producers are also subject to the same agricultural, food safety and other government regulations that apply to non-certified producers.

Purpose of Certification

Organic certification addresses a growing worldwide demand for organic food. It is intended to assure quality and prevent fraud, and to promote commerce. While such certification was not necessary in the early days of the organic movement, when small farmers would sell their produce directly at farmers' markets, as organics have grown in popularity, more and more consumers are purchasing organic food through traditional channels, such as supermarkets. As such, consumers must rely on third-party regulatory certification.

For organic producers, certification identifies suppliers of products approved for use in certified operations. For consumers, "certified organic" serves as a product assurance, similar to "low fat", "100% whole wheat", or "no artificial preservatives".

Certification is essentially aimed at regulating and facilitating the sale of organic products to consumers. Individual certification bodies have their own service marks, which can act as branding to consumers—a certifier may promote the high consumer recognition value of its logo as a marketing advantage to producers. Most UK certification bodies operate organic standards that meet the UK government's minimum requirements. Some certification bodies, such as the Soil Association, certify to higher standards.

Third Party Certification Process

To certify a farm, the farmer is typically required to engage in a number of new activities, in addition to normal farming operations:

- Study the organic standards, which cover in specific detail what is and is not allowed for every aspect of farming, including storage, transport and sale.
- Compliance — farm facilities and production methods must comply with the standards, which may involve modifying facilities, sourcing and changing suppliers, etc.
- Documentation — extensive paperwork is required, detailing farm history and current set-up, and usually including results of soil and water tests.
- Planning — a written annual production plan must be submitted, detailing everything from seed to sale: seed sources,

field and crop locations, fertilization and pest control activities, harvest methods, storage locations, etc.

- Inspection — annual on-farm inspections are required, with a physical tour, examination of records, and an oral interview.
- Fee — an annual inspection/certification fee (currently starting at $400–$2,000/year, in the US and Canada, depending on the agency and the size of the operation).
- Record-keeping — written, day-to-day farming and marketing records, covering all activities, must be available for inspection at any time.

In addition, short-notice or surprise inspections can be made, and specific tests (e.g. soil, water, plant tissue) may be requested.

For first-time farm certification, the soil must meet basic requirements of being free from use of prohibited substances (synthetic chemicals, etc.) for a number of years. A conventional farm must adhere to organic standards for this period, often, two to three years. This is known as being in *transition*. Transitional crops are not considered fully organic.

Certification for operations other than farms is similar. The focus is on ingredients and other inputs, and processing and handling conditions. A transport company would be required to detail the use and maintenance of its vehicles, storage facilities, containers, and so forth. A restaurant would have its premises inspected and its suppliers verified as certified organic.

Participatory Certification

"Participatory Guarantee Systems are locally focused quality assurance systems. They certify producers based on active participation of stakeholders and are built on a foundation of trust, social networks and knowledge exchange" (IFOAM definition, 2008).

Participatory Guarantee Systems (PGS) represent an alternative to third party certification, especially adapted to local markets and short supply chains. They can also complement third party certification with a private label that brings additional guarantees and transparency. PGS enable the direct participation of producers, consumers and other stakeholders in:

- the choice and definition of the standards
- the development and implementation of certification procedures
- the certification decisions

Participatory Guarantee Systems are also referred to as "participatory certification".

Certification and Product Labeling

In some countries, organic standards are formulated and overseen by the government. The United States, the European Union, Canada and Japan have comprehensive organic legislation, and the term "organic" may be used only by certified producers. Being able to put the word "organic" on a food product is a valuable marketing advantage in today's consumer market, but does not guarantee the product is legitimately organic. Certification is intended to protect consumers from misuse of the term, and make buying organics easy. However, the organic labeling made possible by certification itself usually requires explanation. In countries without organic laws, government guidelines may or may not exist, while certification is handled by non-profit organisations and private companies.

Internationally, equivalency negotiations are underway, and some agreements are already in place, to harmonise certification between countries, facilitating international trade. There are also international certification bodies, including members of the International Federation of Organic Agriculture Movements (IFOAM) working on harmonisation efforts. Where formal agreements do not exist between countries, organic product for export is often certified by agencies from the importing countries, who may establish permanent foreign offices for this purpose.

United States of America

In the US, federal organic legislation defines three levels of organics. Products made entirely with certified organic ingredients and methods can be labelled "100% organic". Products with at least 95% organic ingredients can use the word "organic". Both of these categories may also display the USDA organic seal. A third category, containing a minimum of 70% organic ingredients, can be labelled "made with organic ingredients". In addition, products may also display the logo of the certification body that approved them. Products made with less than 70% organic ingredients can not advertise this information to consumers and can only mention this fact in the product's ingredient statement. Similar percentages and labels apply in the EU.

In the U.S., the Organic Foods Production Act of 1990 "requires the Secretary of Agriculture to establish a National List of Allowed and Prohibited Substances which identifies synthetic substances that may be used, and the nonsynthetic substances that cannot be used, in organic production and handling operations."

Also in the U.S., the National Organic Program (NOP), was enacted as federal legislation in October 2002. It restricts the use of the term "organic" to certified organic producers (excepting growers selling under $5,000 a year, who must still comply and submit to a records audit if requested, but do not have to formally apply). Certification is handled by state, non-profit and private agencies that have been approved by the US Department of Agriculture (USDA).

One of the first organisations to carry out organic certification in North America was the California Certified Organic Farmers, founded in 1973.

Canada

In Canada, certification was implemented at the federal level on June 30, 2009. Mandatory certification is required for agricultural products represented as organic in import, export and inter-provincial trade, or that bear the federal organic logo. In Quebec, provincial legislation provides government oversight of organic certification within the province, through the Quebec Accreditation Board (*Conseil D'Accréditation Du Québec*).

Europe

EU countries acquired comprehensive organic legislation with the implementation of the EU-Eco-regulation 1992. Supervision of certification bodies is handled on the national level. In March 2002 the European Commission issued a europeanwide label for organic food however for most of the countries it was not able to replace existing national product labels.

Table 1: European Organic Farmland in 2005

Country	*Area (ha)*	*Percent (%)*
Belgium	22,994	1.7
Czech Republic	254,982	7.2
Cyprus	2	1.1
Denmark	134,129	5.2
Finland	147,587	6.5
France	560,838	2
Germany	807,406	4.7
Greece	288,737	7.2

Hungary	128,576	2
Ireland	34,912	0.8
Italy	1,069,462	8.4
Latvia	118,612	7
Lithuania	64,544	2.3
Luxembourg	3,158 *	2.4
Malta	14	0.1
Netherlands	48,765	2.5
Austria	360,369	11
Poland	82,730 *	2.4
Portugal	233,458	6.3
Sweden	222,268	6.2
Switzerland		11
Slovakia	90,206	4.8
Slovenia	23,499	4.8
Spain	807,569	3.2
United Kingdom	608,952	3.8
EU Total	6,115,465	3.9

Source: "Eurostat press release 80/2007"

The European organic food label has been mandatory throughout the EU since July 2010. Originally it was planned to replace the old national organic labels but it was finally decided to allow for a transition period where both the EU label and the national labels may be used to mark organic food. The new logo is a green rectangle that shows twelve stars (from the European flag) placed such that they form the shape of a leaf in the wind. Unlike earlier labels no words are presented on the label lifting the requirement for translations referring to organic food certification.

In the United Kingdom, organic certification is handled by a number of organisations, regulated by The Department for Environment, Food and Rural Affairs (DEFRA), of which the largest are the Soil Association and Organic Farmers and Growers.

In Sweden, organic certification is handled by the organisation with members such as farmers, processors, trade and also consumer, environmental and animal welfare interests.

In Ireland, organic certification is available from the Irish Organic Farmers and Growers Association

In Greece, organic certification is available from eight (8) organisations approved by EU. The major of them are BIOHELLAS and the DIO.

In France, organic certification was introduced in 1985. It has established a green-white logo of "AB - agriculture biologique". The certification for the AB label fulfills the EU regulations for organic food. The certification process is overseen a public institute ("Agence française pour le développement et la promotion de l'agriculture biologique" usually shortended to "Agence bio") established in November 2001. The actual certification authorities include a number of different institutes like Aclave, Agrocert, Ecocert SA, Qualité France SA, Ulase, SGS ICS.

In Belgium, a similar process as in France is being used where certification of the "Biogarantie" label is overseen by the public "association sans but lucratif" (ASBL) administration. This administration does exist since 27. June 1921 and it was reformed on 2. May 2002 to take over the new responsibilities of the label certification.

In Germany the national label was introduced in September 2001 following in the footsteps of the political campaign of "Agrarwende" (*agricultural major shift*) led by minister Renate Künast of the Greens party. This campaign was started after the mad-cow disease epidemic in 2000. The effects on farming are still challenged by other political parties. The national "Bio"-label in its hexagon green-black-white shape has gained wide popularity - in 2007 there were 2431 companies having certified 41.708 products. The popularity of the label is extending to neighbouring countries like Austria, Switzerland and France.

In the German-speaking countries there have been older non-government organisations that had issued labels for organic food long before the advent of the EU organic food regulations. Their labels are still used widely as they significantly exceed the requirements of the EU regulations. An organic food label like "demeter" from Demeter International has been in use since 1928 and this label is still regarded as providing the highest standards for organic food in the world. Other active NGOs include Biokreis, Bioland, Biopark, Ecoland, Ecovin, Gäa e.V. and Naturland.

The farmland converted to produce certified organic food has seen a signification evolution in the EU15 rising from 1.8% in 1998 to 4.1% in 2005. For the current EU25 however the statistics do report an overall percentage of just 1.5% as of 2005. Other than the percentage of farmland the statistics show a much larger percentage of organic food in terms of turnover reaching 10% in France and 14% in Germany. Vegetables, fruits, milk and eggs have reached a percentage of 21% to be certified organic food in France. Numbers of 2010 show that 5.4% of farmland in Germany have been converted to produced certified organic food, Switzerland reaches 10.4% and Austria rises to 11.7% of the farmland. Non-EU countries have widely adopted the European certification regulations for organic food, to increase export chances to EU countries.

In Other Countries

In Japan, the Japanese Agricultural Standard (JAS) was fully implemented as law in April 2001. This was revised in November 2005 and all JAS certifiers were required to be re-accredited by the Ministry of Agriculture.

In Australia, the Australian Quarantine and Inspection Service (AQIS) is the controlling body for organic certification because there are no domestic standards for organic produce within Australia. Currently the government only becomes involved with organic certification at export, meaning AQIS is the default certification agency. Although there is no system for monitoring the labeling of organic produce sold within Australia, this primarily affects the retail public. Commercial buyers for whom this is an issue have simply taken the export system as a de facto standard and are willing to pay premium prices for produce from growers certified under the National schemes. Although it is voluntary for domestic product to adhere to the AQIS standard when claiming the product as "organic", it is mandatory for any product making claims to certified organic, to be verifiable as it being certified organic. As of 2010, there are seven AQIS-approved certifying organisations, they are AUS-QUAL Limited, Australian Certified Organic, Bio-Dynamic Research Institute, NASAA Certified Organic, Safe Food Productions Queensland, Tasmanian Organic-dynamic Producers these organisations are authorised to issue Organic Produce Certificates on behalf of AQIS. In 2004 there were 2345 certified operators in Australia. There are some private organic standards that do not meet the export act and as such these products and/or ingredients are not eligible to be exported from Australia as they are not in compliance with the Organic Orders and/or related export legislation.

The largest importer of Australia's organic produce (by weight) is Japan (33.59%), followed by the UK (17.51%), France (10.51%), and

New Zealand (10.21%). The largest certifier of organic products is Australian Certified Organic, which is a subsidiary of Biological Farmers Australia, the largest organic farmers' collective in the country. There is another organic farmer's collective in Australia called the Organic Federation of Australia.

In India, APEDA regulates the certification of organic products as per National Standards for Organic Production. "The NPOP standards for production and accreditation system have been recognised by European Commission and Switzerland as equivalent to their country standards. Similarly, USDA has recognised NPOP conformity assessment procedures of accreditation as equivalent to that of US. With these recognitions, Indian organic products duly certified by the accredited certification bodies of India are accepted by the importing countries."

In China, the China Green Food Development Centre awards two Standards: A and AA; while the former standard does permit some use of synthetic agricultural chemicals, the latter is more stringent.

Certification Issues

Organic certification is not without its critics. Some of the staunchest opponents of chemical-based farming and factory farming practices also oppose formal certification. They see it as a way to drive independent organic farmers out of business, and to undermine the quality of organic food. Other organisations such as the Organic Trade Association work within the organic community to foster awareness of legislative and other related issues, and enable the influence and participation of organic proponents.

Obstacle to Small Independents

Originally, in the 1960s through the 1980s, the organic food industry was composed of mainly small, independent farmers, selling locally. Organic "certification" was a matter of trust, based on a direct relationship between farmer and consumer. Critics view regulatory certification as a potential barrier to entry for small producers, by burdening them with increased costs, paperwork, and bureaucracy.

Manipulation of Regulations

Critics of formal certification also fear an erosion of organic standards. Provided with a legal framework within which to operate, lobbyists can push for amendments and exceptions favourable to large-scale production, resulting in "legally organic" products produced in ways similar to current conventional food. Combined with the fact that organic products are now sold predominantly through high volume distribution channels such as supermarkets, the concern is that the market is evolving to favour the biggest producers, and this could result in the small organic farmer being squeezed out.

Manipulation of certification regulations as a way to mislead or outright dupe the public is a very real concern. Some examples are creating exceptions (allowing non-organic inputs to be used without loss of certification status) and creative interpretation of standards to meet the letter, but not the intention, of particular rules.

For example, a complaint filed with the USDA in February 2004 against Bayliss Ranch, a food ingredient producer and its certifying agent, charged that tap water had been certified organic, and advertised for use in a variety of water-based body care and food products, in order to label them "organic" under US law. Steam-distilled plant extracts, consisting mainly of tap water introduced during the distilling process, were certified organic, and promoted as an organic base that could then be used in a claim of organic content.

The case was dismissed by the USDA, as the products had been actually used only in personal care products, over which the department at the time extended no labeling control. The company subsequently adjusted its marketing by removing reference to use of the extracts in food products. Several months later, the USDA extended its organic labeling to personal care products; this complaint has not been refiled.

In December 2005, the 2006 agricultural appropriations bill was passed with a rider allowing 38 synthetic ingredients to be used in organic foods. Among the ingredients are food colourings, starches, sausage and hot-dog casings, hops, fish oil, chipotle chili pepper, and gelatin. This allowed Anheuser-Busch in 2007 to have its Wild Hop Lager certified organic "even though [it] uses hops grown with chemical fertilizers and sprayed with pesticides."

Misrepresentation of the Term Organic

The word *organic* is central to the certification (and organic food marketing) process, and this is also questioned by some. Where organic laws exist, producers cannot use the term legally without certification. To bypass this legal requirement for certification, various alternative certification approaches, using currently undefined terms like "authentic" and "natural", are emerging. In the US, motivated by the cost and legal requirements of certification (as of Oct. 2002), the private farmer-to-farmer association, Certified Naturally Grown, offers a "non-profit alternative eco-labelling program for small farms that grow using USDA Organic methods but are not a part of the USDA Certified Organic program."

In the UK, the interests of smaller-scale growers who use 'natural' growing methods are represented by the Wholesome Food Association, which issues a symbol based largely on trust and peer-to-peer inspection.

A related concern holds that certification is replacing consumer education, and this goes against the essential, holistic nature of organic farming. By reducing complex issues and regulations to a simple, convenient *certified organic* label, consumers may more easily ignore the principles and practices behind organics, leaving the definition of organic farming and organic food open to manipulation.

Standards

Standards regulate production methods and in some cases final output for organic agriculture. Standards may be voluntary or legislated. As early as the 1970s private associations certified organic producers. In the 1980s, governments began to produce organic production guidelines. In the 1990s, a trend toward legislated standards began, most notably with the 1991 EU-Eco-regulation developed for European Union, which set standards for 12 countries, and a 1993 UK program.

The EU's program was followed by a Japanese program in 2001, and in 2002 the U.S. created the National Organic Program (NOP). As of 2007 over 60 countries regulate organic farming (IFOAM 2007:11). In 2005 IFOAM created the Principles of Organic Agriculture, an international guideline for certification criteria. Typically the agencies accredit certification groups rather than individual farms.

Organic production materials used in and foods are tested independently by the Organic Materials Review Institute.

Composting

Under USDA organic standards, manure must be subjected to proper thermophilic composting and allowed to reach a sterilizing temperature. If raw animal manure is used, 120 days must pass before the crop is harvested if the final product comes into direct contact with the soil. For products which do not come into direct contact with soil, 90 days must pass prior to harvest.

Economics

The economics of organic farming, a subfield of agricultural economics, encompasses the entire process and effects of organic farming in terms of human society, including social costs, opportunity costs, unintended consequences, information asymmetries, and economies of scale. Although the scope of economics is broad, agricultural economics tends to focus on maximising yields and efficiency at the farm level. Economics takes an anthropocentric approach to the value of the natural world: biodiversity, for example, is considered beneficial only to the extent that it is valued by people and increases profits. Some entities such as the European Union subsidize organic farming, in large part because these countries want to account for the externalities of reduced water use, reduced water contamination, reduced soil erosion, reduced carbon emissions, increased biodiversity, and assorted other benefits that result from organic farming.

Traditional organic farming is labour and knowledge-intensive whereas conventional farming is capital-intensive, requiring more energy and manufactured inputs.

Organic farmers in California have cited marketing as their greatest obstacle.

Geographic Producer Distribution

The markets for organic products are strongest in North America and Europe, which as of 2001 are estimated to have $6 and $8 billion respectively of the $20 billion global market (Lotter 2003:6). As of 2007 Australasia has 39% of the total organic farmland, including Australia's 1,180,000 hectares (2,900,000 acres) but 97 percent of this land is sprawling rangeland (2007:35). US sales are 20x as much. (2003:7). Europe farms 23 percent of global organic farmland (6.9 million hectares), followed by Latin America with 19 percent (5.8 million hectares). Asia has 9.5 percent while North America has 7.2 percent. Africa has 3 percent.

Besides Australia, the countries with the most organic farmland are Argentina (3.1 million hectares), China (2.3 million hectares), and the United States (1.6 million hectares). Much of Argentina's organic farmland is pasture, like that of Australia (2007:42). Italy, Spain, Germany, Brazil (the world's largest agricultural exporter), Uruguay, and the UK follow the United States in the amount of organic land (2007:26).

Growth

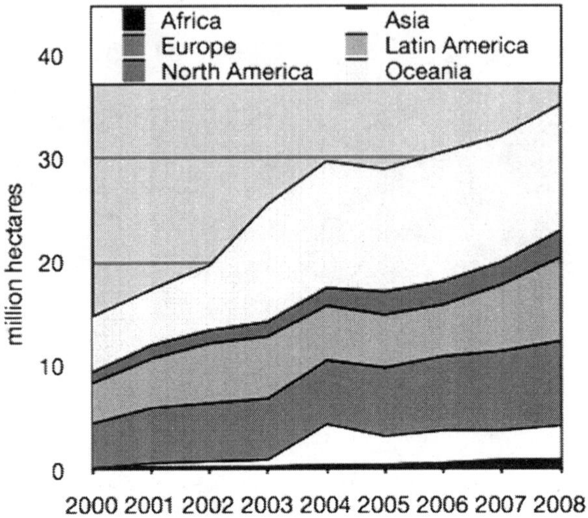

Figure 6: Organic farmland by world region (2000-2008)

As of 2001, the estimated market value of certified organic products was estimated to be $20 billion. By 2002 this was $23 billion and by 2007 more than $46 billion.

In recent years both Europe (2007: 7.8 million hectares, European Union: 7.2 million hectares) and North America (2007: 2.2 million hectares) have experienced strong growth in organic farmland. In the EU it grew by 21% in the period 2005 to 2008. However, this growth has occurred under different conditions. While the European Union has shifted agricultural subsidies to organic farmers due to perceived environmental benefits, the United States has not, continuing to subsidize some but not all traditional commercial crops, such as corn and sugar. As a result of this policy difference, as of 2008 4.1% percent of European Union farmland was organically managed compared to the 0.6 percent in the U.S.

IFOAM's most recent edition of *The World of Organic Agriculture: Statistics and Emerging Trends 2009* lists the countries which had the most hectares in 2007. The country with the most organic land is Australia with more than 12 million hectares, followed by Argentina, Brazil and the US. In total 32.2 million hectares were under organic management in 2007. For 1999 11 million hectares of organically managed land are reported.

As organic farming becomes a major commercial force in agriculture, it is likely to gain increasing impact on national agricultural policies and confront some of the scaling challenges faced by conventional agriculture.

Productivity and Profitability

Various studies find that versus conventional agriculture, organic crops yielded 91%, or 95-100%, along with 50% lower expenditure on fertilizer and energy, and 97% less pesticides, or 100% for corn and soybean, consuming less energy and zero pesticides. The results were attributed to lower yields in average and good years but higher yields during drought years.

A 2007 study compiling research from 293 different comparisons into a single study to assess the overall efficiency of the two agricultural systems has concluded that:

> *...organic methods could produce enough food on a global per capita basis to sustain the current human population, and potentially an even larger population, without increasing the agricultural land base.*

Converted organic farms have lower pre-harvest yields than their conventional counterparts in developed countries (92%) but higher than their low-intensity counterparts in developing countries (132%). This is due to relatively lower adoption of fertilizers and pesticides in the developing world compared to the intensive farming of the developed world. Organic farms withstand severe weather conditions better than conventional farms, sometimes yielding 70-90% more than conventional farms during droughts. Organic farms are more profitable in the drier states of the United States, likely due to their superior drought performance. Organic farms survive hurricane damage much better, retaining 20 to 40% more topsoil and smaller economic losses at highly significant levels than their neighbours.

Contrary to widespread belief, organic farming can build up soil organic matter better than conventional no-till farming, which suggests long-term yield benefits from organic farming. An 18-year study of organic methods on nutrient-depleted soil, concluded that conventional methods were superior for soil fertility and yield in a cold-temperate climate, arguing that much of the benefits from organic farming are derived from imported materials which could not be regarded as "self-sustaining".

Profitability

The decreased cost of synthetic fertilizer and pesticide inputs, along with the higher prices that consumers pay for organic produce, contribute to increased profits. Organic farms have been consistently found to be as or more profitable than conventional farms. Without the price premium, profitability is mixed. Organic production was more profitable in Wisconsin, given price premiums.

Sustainability (African Case)

In 2008 the United Nations Environmental Programme (UNEP) and the United Nations Conference on Trade and Development (UNCTAD) stated that "organic agriculture can be more conducive to food security in Africa than most conventional production systems, and that it is more likely to be sustainable in the long-term" and that "yields had more than doubled where organic, or near-organic practices had been used" and that soil fertility and drought resistance improved.

Employment Impact

Organic methods often require more labour than traditional farming, therefore it provides rural jobs.

Externalities

Agriculture imposes negative externalities (uncompensated costs) upon society through land and other resource use, biodiversity loss, erosion, pesticides, nutrient runoff, water usage, subsidy payments and assorted other problems. Positive externalities include self-reliance, entrepreneurship, respect for nature, and air quality. Organic methods reduce some of these costs. In 2000 uncompensated costs for 1996 reached 2,343 million British pounds or 208 pounds per hectare. In 2005 in the USA concluded that cropland costs the economy approximately 5 to 16 billion dollars ($30 to $96 per hectare), while livestock production costs 714 million dollars. Both studies recommended reducing externalities. The 2000 review included reported pesticide poisonings but did not include speculative chronic health effects of pesticides, and the 2004 review relied on a 1992 estimate of the total impact of pesticides.

It has been proposed that organic agriculture can reduce the level of some negative externalities from (conventional) agriculture. Whether the benefits are private or public depends upon the division of property rights.

Chapter 2

Motivations for Organic Agriculture

Within the food industry, defining the benefits of organic food is largely left to word of mouth, media coverage, and the promotional efforts of organic advocates. Major food and beverage corporations have rapidly moved to acquire significant stake in both fresh and processed organic products. Still, the specific sales points of "organics" go largely unmentioned on product packaging and in mainstream media advertising.

Claims of improved food quality are regularly used in conventional food marketing, with "low fat", "low sodium", "whole grain", "high fibre", "vitamin enriched", "no trans-fat" and other commonly advertised benefits. By contrast, "certified organic" is generally left to stand on its own as self-explanatory, assisted only by general terms like "natural".

Consumer surveys have consistently identified food quality as the main reason for purchasing organic food. Higher nutritional value, no toxic residues from pesticides, and better taste are often cited, as is the positive impact of organic production on the environment. Whether organic food actually delivers on these desires and beliefs is controversial and the subject of scientifically inconclusive debate. The debate centres on a variety of specific and supposedly demonstrable characteristics which proponents claim make organic food superior to the product of conventional farming and processing.

Food Safety

Organic food proponents express concern over the potential negative effects of various chemical cultivation methods and genetic modification techniques used in modern conventional agriculture. The effect of pesticide residues from crop spraying, the presence of veterinary drugs in meat products, and the relatively unknown long term impacts of genetically modified varieties and breeds are all encompassed. Organic food is seen as avoiding relieving these concerns by prohibiting such practices. At present, there are no definitive scientific conclusions on any of these matters; individual studies are cited on both sides of the debate.

Chemical Contamination

Organic food proponents cite the existence of reduced levels of pesticides and herbicides as a way to reduce the long term risk of chemical consumption. The degree of risk posed by pesticide residues remains uncertain. Pesticide use in conventional food products is heavily regulated, with established, research-based maximum residue levels (MRLs) below which residues are considered safe for human consumption. Also, many pesticides are not cumulative in the body, and are regularly eliminated. Notable exceptions include heavy metals such as lead or mercury which are sometimes found in foodstuffs in countries which have lax food production standards. The U.S. and most of Europe prohibit the use of inorganic compounds containing heavy metals in any type of agriculture including conventional. Organic farming standards do not allow the use of synthetic pesticides, but they do allow the use of certain so-called natural pesticides, such as those derived from plants. The most common organic pesticides, accepted for restricted use by most organic standards, include Bt, pyrethrum, and rotenone. Some organic pesticides, such as rotenone, have high toxicity to fish and aquatic creatures with some toxicity to mammals including humans.

A study published by the National Research Council in 1993 determined that for infants and children, the major source of exposure to pesticides is through diet. A recent study in 2006 measured the levels of organophosphorus pesticide exposure in 23 school children before and after replacing their diet with organic food. In this study it was found that levels of organophosphorus pesticide exposure dropped dramatically and immediately when the children switched to an organic diet.

One of the major differences in the production of organically produced food from traditional techniques is its reliance on so-called natural methods of providing soil nutrients. While both methods rely on nitrates to provide nitrogen to the soil, organic agriculture gets the majority of its nitrates from manure and composting sources. These sources of nitrogen have a slower transfer of nitrogen to the soil spreading out the release of nitrogen over the growing season and helping to reduce the leeching of nitrates into water sources where they can cause health problems. However, the slower transfer can also increase nitrate run-off. Manure and compost also contain the other macronutrients, phosphate and potassium, and micronutrients. While neither chemical or organic applications of nutrients to the soil provide the ideal situation, on small to medium scale operations, manure provides the benefit of reduced soil erosion and less leeching of nitrates into the soil. Regardless of the benefits, the USDA argues that the application of man-made nutrients will always be required to sustain a global agriculture system that feeds the entire world. Others, however, say that organic farming need not result in reduced production.

Organic methods of fertilization are not necessarily free from risk. However, organic animal manure, typically that of cattle, is manure from animals that eat mainly hay and other organic, primarily non-grain materials. This is seen as a way to reduce the amount of E. coli bacteria present, and the feces of organically-raised cattle have only 1% of the E. coli present in non-organic manure. It should also be noted that many organic farmers consider using manure directly as fertilizer to be an unsound practice, and instead the manure should be composted first.

Hormonal Contamination

Organic proponents cite evidence that some chemicals used in conventional farming, including pesticides and herbicides, mimic hormones - usually estrogen - when inside a person. They claim that this is significant even at the minute levels that the average person is exposed to. The US government states that these chemicals are safe when used correctly, but proponents claim such tests are only done on healthy adults - and that children and fetuses might be at risk to even small amounts of these chemicals.

In Australia, the Government sponsored Australian Total Diet Survey measures pesticide residues found in typical Australian diets. The 2004 survey found all estimated dietary exposures to pesticide residues were below 16% of the respective Acceptable daily intakes and therefore all exposures are well within the applicable health standards.

Transgenic Contamination

The health risks surrounding genetically modified foods remain highly contentious. Other issues surrounding Genetically modified organisms (GMOs) may also concern consumers, such as the ownership of biological intellectual property by corporations, and reduction in crop varieties.

Certified organic foods are not substantially genetically modified. GMO contamination of organic crops may unintentionally occur primarily through either pollen drift and/or the handling of raw organic agricultural products. There exist estimates that pollen of some crops (e.g. canola) can travel more than 5 kilometres per year.

In the United States, genetic modification is a prohibited/excluded method in organic production as defined in the National Organic Program (NOP). The USDA does not require GMO testing for all crops marketed or labelled as organic and therefore does not technically regulate GMO traces for such. The USDA does verify, through Accredited Certifying Agents, the composition and percentage of organic ingredients for all food products labelled or marketed as certified organic in the United States. The NOP does not prescribe that an Accredited Certifying Agent (ACA) revoke or suspend certification of an organic farm if accidental GMO pollen drift occurs. But all ACA's have the authority to test soils and/or crops, if necessary, for GMO's and other prohibited substances and methods. If organic certification is continued for an organic farm affected by GMO contamination (almost always through pollen drift) the farmer does have an opportunity to market most future organic crops. However, the farmers' ability to market the GMO contaminated organic crop is minimised substantially and the organic market premium is lost.

In most European countries, certification rules are much stricter. Basically, any confirmed detection of transgenic plant, seed or feed can result in a loss of organic status and consequent substantial economic losses for the farmer.

Nutritional Value

Some organic advocates claim that organic food is more nutritious. Increased soil quality, greater attention to quality, and selection of crop varieties for nutrition and taste instead of size, appearance, and shipping characteristics are claimed to be reasons for higher nutrient density of organic foods.

Organic advocates claim that organically grown potatoes, oranges, and leafy vegetables have more vitamin C than conventionally grown products. Phenolic compounds are also found in significantly higher concentrations in organic foods, and these may provide antioxidant protection against heart disease and cancer.

Still isolated bits of research suggest that conventional agricultural practices are degrading food quality. A study in the *Journal of the American College of Nutrition* in 2004, entitled *Changes in USDA*

Food Composition Data for 43 Garden Crops, 1950 to 1999, compared vegetables analysed in 1950 and in 1999, and found noticeable decreases in six of 13 nutrients examined (the six were: protein, calcium, phosphorus, iron, riboflavin and ascorbic acid). Percentage reductions ranged from 6% for protein to 38% of riboflavin, although when evaluated on a per-food or per-nutrient level, usually no distinguishable changes were found. Reductions in calcium, phosphorus, iron and ascorbic acid were also found. The authors suggested that the differences probably reflect changes in cultivated varieties between 1950 and 1999, in which there may have been trade-offs between yield and nutrient content.

Whether organic foodstuffs have a higher nutrient content is still debatable. Studies have shown no clear, consistent results, and those that have suffer from significant experimental design flaws, according to the United Nation's Food and Agriculture Organisation (FAO). Further, the FAO found that in some crops, such as wheat, there appears to be a trade-off: in conventionally farmed wheat the levels of protein are higher, but the lower levels in organic materials are offset by gains in alpha-amylase and sugar contents.

Taste

Many claim that organic food tastes better. This is primarily referred to regarding fresh food. It is possible that organic food tastes better simply because it is fresher. Because organic farms tend to be smaller, they often sell their products closer to the point of harvest. Thus, organic fruits and vegetables taste more "farm fresh" than comparable conventional produce. However, organic foods might also have more flavour because organic farmers often breed with taste instead of marketability as the primary factor. Conventional tomatoes, for example, are often bred to be perfectly red and round, to match the ideal appearance of a tomato. They are also bred to resist damage in transport and storage, for a longer shelf-life. This means that taste is an attribute that has a lower priority. In addition to crop diversity and selection practices, organic farming emphasizes soil nutrition, which can positively influence the taste of the food. Tests by the United Nations FAO demonstrated that some apples, specifically the "Golden Delicious" variety, have higher flavonoid counts when grown organically. This suggests that they do have more flavour.

Some foods, such as bananas, are picked when unripe, then artificially induced to ripen using a chemical (such as propylene or ethylene) while in transit, possibly producing a different taste. The issue of ethylene use in organic food production is contentious; opponents claiming that its use only benefits large companies, and opens the door to weaker organic standards.

Environmental Impact

Every food purchase supports the system that delivers it, and if large-scale chemical production methods are damaging to the environment, then purchasing these foods supports this damage. A main goal of organic farming is minimising impact to the environment.

Sustainability

Proponents of organic farming say that "conventional" farming is unsustainable, because it relies on artificial inputs (synthetic fertilizers and other chemicals, machinery, etc.) that ultimately requires energy in the form of fossil fuels, and because the land is degraded through soil erosion, salinization, and other processes that eventually render the soil infertile. Many claim that without cheap fossil fuels and government subsidies, conventional agriculture would not be possible, and that despite technological advancements, there will eventually be an agricultural crisis as a result of depleted soil. The cultivation of monocultures, many acres planted with the same crop year after year, increases susceptibility to pests and diseases and depletes the soil, while eliminating most native flora and fauna.

In contrast, organic farming often utilises intercropping, crop rotation, fallow periods, and integrated pest management to promote biodiversity and preserve the health of the soil while minimising the risk of diseases. The main goal of organic farming is sustainability, so organic farms seek to minimise dependence on outside resources and be self-sufficient.

Pollution

Modern agricultural practices often result in large amounts of nitrogen runoff from the heavy use of fertilizer, which pollutes watersheds. In addition to posing a threat to human health and disrupting aquatic ecosystems, this sometimes results in algal blooms which deplete the water of oxygen resulting in fishkills. Pesticide runoff also causes many problems.

Transportation

Claims that eating organic food is better for the environment are, however, frustrated by the fact that most of the organic food sold today travels the same great distances as conventional food. A UK study published in 2005 in the *Journal of Food Policy* found that maximum environmental benefit would result from purchasing food produced within a 12-mile radius. Therefore, buying local food that is not organic could be environmentally "better" than buying organic food that has travelled hundreds or thousands of miles. Many small organic farms, however, sell much or all of their produce locally.

Energy Efficiency

Studies have shown organic farms to be more energy efficient than their conventional counterparts. One of these studies was done with apple farms in the state of Washington. In that study, the organic farms were found to be at least 7% more energy efficient. Although some critics of organic farms cite evidence that organic farms produce less yield than conventional farms, this study also found a much more substantial decrease in resources used. One prominent 21-year Swiss study found an average 20% lower organic yields over conventional methods. However, that came with consumption of 50% less fertilizer, and 97% less pesticide.

In comparing yields, a US survey published in 2001 analysed 150 growing seasons of data on various crops and concluded that organic yields were 95-100% of conventional yields. Because organic farms don't use toxic pesticides and herbicides, there is more biodiversity in the soil. Besides higher soil quality - more life in the soil allows for higher water retention. This helps increase yields for organic farms in drought years where there is less rain. During drought years, organic farms have been found to have yields 20-40% higher than conventional farms.

Large Scale Organic Farms

Many advocates of organic farming view large scale, corporate owned "organic farms" as being against the spirit of organic farming, since they tend to use unsustainable practices similar to conventional farms.

Permaculture

Permaculture is an approach to designing human settlements and agricultural systems that is modelled on the relationships found in nature. It is based on the ecology of how things interrelate rather than on the strictly biological concerns that form the foundation of modern agriculture. Permaculture aims to create stable, productive systems that provide for human needs; it's a system of design where each element supports and feeds other elements, ultimately aiming at systems that are virtually self-sustaining and into which humans fit as an integral part.

Permaculture as a systematic method was developed by Australians Bill Mollison and David Holmgren during the 1970s. The word "permaculture" originally referred to "permanent agriculture", but was expanded to also stand for "permanent culture" as it was seen that social aspects were integral to a truly sustainable system. Mollison has described permaculture as "*a philosophy of working with, rather than against nature; of protracted and thoughtful observation rather than protracted and thoughtless labour; and of looking at plants and animals in all their functions, rather than treating any area as a single project system.*"

Permaculture draws from several other disciplines including organic farming, agroforestry, sustainable development, and applied ecology. "*The primary agenda of the movement has been to assist people to become more self reliant through the design and development of productive and sustainable gardens and farms. The design principles which are the conceptual foundation of permaculture were derived from the science of systems ecology and study of pre-industrial examples of sustainable land use.*"

History

Early Influences

Franklin Hiram King coined the term *permanent agriculture* in his classic book from 1911, *Farmers of Forty Centuries: Or Permanent Agriculture in China, Korea and Japan.* In this context, permanent agriculture is understood as agriculture that can be sustained indefinitely.

In 1929, Joseph Russell Smith took up the term as the subtitle for *Tree Crops: A Permanent Agriculture*, a book in which he summed up his long experience experimenting with fruits and nuts as crops for human food and animal feed. Smith saw the world as an interrelated whole and suggested mixed systems of trees and crops underneath. The definition of permanent agriculture as that which can be sustained indefinitely was supported by Australian P. A. Yeomans in his 1973 book "*Water for Every Farm*". Yeoman introduced an observation-based approach to land use in Australia in the 1940s; and the keyline design as a way of managing the supply and distribution of water in the 1950s.

Stewart Brand's works was an early influence noted by Holmgren. Brand is an advocate of systems thinking and was a major figure in the counterculture movement of the Sixties and Seventies. Systems thinking proposes to view systems in a holistic manner. Other early influences include Ruth Stout and Esther Deans, who pioneered "no-dig gardening methods", and Masanobu Fukuoka who, in the late 1930s in Japan, began advocating no-till orchards, gardens and natural philosophy.

The first recorded modern practice of permaculture as a systematic method was possibly by Austrian farmer Sepp Holzer in the 1960s. Coming from a line of farmers, Holzer took over his parents' mountain farm business in 1962 and pioneered the use of ecological farming, or permaculture, techniques at high altitudes (1100 to 1500 metres above sea level) after being unsuccessful with regular farming methods.

Mollison and Holmgren

In the mid 1970s, Bill Mollison and David Holmgren started developing ideas about stable agricultural systems on the southern Australian island state of Tasmania. This was a result of their perception of a rapidly growing usage of industrial-agricultural methods. In their view these methods were poisoning the land and water, reducing biodiversity, and removing billions of tons of topsoil from previously fertile landscapes. A design approach called "permaculture" was their response and was first made public with the publication of their book "*Permaculture One*" in 1978.

By the early 1980s, the concept had broadened from agricultural systems design towards complete, sustainable human habitats. After *Permaculture One*, Mollison further refined and developed the ideas by designing hundreds of permaculture sites and writing more detailed books, notably "*Permaculture: A Designers Manual*". Mollison lectured in over 80 countries and taught his two-week Permaculture Design Course (PDC) to many hundreds of students.

By the mid 1980s, many of the students had become successful practitioners and had themselves begun teaching the techniques they had learned. In a short period of time permaculture groups, projects, associations, and institutes were established in over one hundred countries. In 1991 a four-part Television documentary by ABC productions called "The Global Gardener" showed permaculture applied to a range of worldwide situations, bringing the concept to a much broader public.

Core Values

At the heart of permaculture design and practice is a fundamental set of core values or ethics which remain constant whatever a person's situation. These ethics are often summarised as;

- *Earthcare*: recognising Earth as the source of all life and recognising that mankind is part of Earth, not apart from it.
- *Peoplecare*: supporting and helping each other live in ways that harm neither ourselves nor the planet and develop healthy societies.
- *Fairshare* : using Earth's limited natural resources in ways that are equitable and wise.

Sustainable Designs

Permaculture design emphasizes patterns of landscape, function, and species assemblies. It asks the question, "Where does this (element) go? How can it be placed for the maximum benefit of the system? Permaculture draws on the practical application of ecological theory to analyse the characteristics of a farm, garden or home site. Each element of a design is carefully analysed in terms of its needs, outputs, and properties. Design elements are then assembled in relation to one another so that the products of one element feed the needs of adjacent elements. Synergy (two or more things functioning together to produce a result not independently obtainable) between design elements is achieved while minimising waste and the demand for human labour or the input of energy. Exemplary permaculture designs evolve over time, and can become extremely complex systems that produce a high density of food and materials with minimal input.

The key to the permaculture design model is that useful connections are made between components in the final design. The focus is not on these elements themselves, but rather on the relationships created among them by the way they are placed together; the whole becoming greater than the sum of its parts. Permaculture is also about careful and contemplative observation of nature and natural systems, and of recognising universal patterns and principles, then learning to apply these to one's own circumstances.

Permaculture is a form of polyculture agriculture. Polyculture is agriculture using multiple crops in the same space, in imitation of the diversity of natural ecosystems, and avoiding large stands of single crops, or monoculture. Perennial plants are often used in permaculture design. As they do not need to be planted every year they require less maintenance and fertilizers. They are especially important in the outer zones and in layered systems.

Design Tools

Modern permaculture is a system design tool. It is a way of:

- looking at a whole system or problem;
- observing how the parts relate;
- planning to mend inefficient systems by applying ideas learned from long-term sustainable working systems;
- seeing connections between key parts.

Holmgren's 12 Design Principles

The core of permaculture has always been in supplying a design toolkit for human habitation. This toolkit helps the designer to model a final design based on an observation of how ecosystems interact.

1. *Observe and Interact*: By taking time to engage with nature we can design solutions that suit our particular situation.
2. *Catch and Store Energy*: By developing systems that collect resources at peak abundance, we can use them in times of need.
3. *Obtain a Yield*: Ensure that you are getting truly useful rewards as part of the work that you are doing.
4. *Apply Self-regulation and Accept Feedback*: We need to discourage inappropriate activity to ensure that systems can continue to function well.
5. *Use and Value Renewable Resources and Services*: Make the best use of nature's abundance to reduce our consumptive behaviour and dependence on non-renewable resources.
6. *Produce no waste*: By valuing and making use of all the resources that are available to us, nothing goes to waste.
7. *Design from Patterns to Details*: By stepping back, we can observe patterns in nature and society. These can form the backbone of our designs, with the details filled in as we go.
8. *Integrate Rather than Segregate*: By putting the right things in the right place, relationships develop between those things and they work together to support each other.

9. *Use small and slow solutions*: Small and slow systems are easier to maintain than big ones, making better use of local resources and producing more sustainable outcomes.
10. *Use and value diversity*: Diversity reduces vulnerability to a variety of threats and takes advantage of the unique nature of the environment in which it resides.
11. *Use edges and value the marginal*: The interface between things is where the most interesting events take place. These are often the most valuable, diverse and productive elements in the system.
12. *Creatively use and respond to change*: We can have a positive impact on inevitable change by carefully observing, and then intervening at the right time.

Patterns

Permaculture design focuses heavily upon natural patterns. All things, even the wind, the waves and the earth moving around the Sun, form patterns. In pattern application, permaculture designers are encouraged to develop an awareness of the patterns that exist in nature (and how these function) and how patterns can be utilised to satisfy the specific design needs of a specific site. "The application of pattern on a design site involves the designer recognising the shape and potential to fit these patterns or combinations of patterns comfortably onto the landscape".

Edge Effect

The edge effect in ecology is the effect of the juxtaposition or placing side by side of contrasting environments on an ecosystem. Permaculturists maintain that, where vastly differing systems meet, there is an intense area of productivity and useful connections. An example of this is the coast; where the land and the sea meet there is a particularly rich area that meets a disproportionate percentage of human and animal needs. So this idea is played out in permacultural designs by using spirals in the herb garden or creating ponds that have wavy undulating shorelines rather than a simple circle or oval (thereby increasing the amount of edge for a given area). Edges between woodland and open areas have been claimed to be the most productive.

Guilds

A guild is any group of species that exploit the same resources, often in related ways. Guilds are groups of plants, animals, insects, etc. that work well together. Some plants may be grown for food production, some to attract beneficial insects, and others to repel harmful insects. When grouped together these plants form a guild.

The end goal is to have a garden that requires little or no ongoing human maintenance or resource inputs. Everything that is added into the system either improves the system or degrades the system. Finding those plants or animals that complement each other, is the first step in designing a useful system. The Three Sisters of maize, squash and beans is a well known example. Guilds can be thought of as an extension of companion planting.

Zones

Zones are a way of organising design elements in a human environment on the basis of the frequency of human use and plant or animal needs. Frequently manipulated or harvested elements of the design are located close to the house in zones 1 and 2. Less frequently used or manipulated elements, and elements that benefit from isolation (such as wild species) are farther away. Zones is about positioning things appropriately. Zones are numbered from 0 to 5.

Zone 0: The house, or home center. Here permaculture principles would be applied in terms of aiming to reduce energy and water needs, harnessing natural resources such as sunlight, and generally creating a harmonious, sustainable environment in which to live and work.

Zone 1: The zone nearest to the house, the location for those elements in the system that require frequent attention, or that need to be visited often, such as salad crops, herb plants, soft fruit like strawberries or raspberries, greenhouse and cold frames, propagation area, worm compost bin for kitchen waste, and so on. Raised beds are often used in zone 1 in urban areas.

Zone 2: This area is used for siting perennial plants that require less frequent maintenance, such as occasional weed control or pruning, including currant bushes and orchards. This would also be a good place for beehives, larger scale composting bins, and so on.

Zone 3: The area where maincrops are grown, both for domestic use and for trade purposes. After establishment, care and maintenance required are fairly minimal (provided mulches and similar things are used), such as watering or weed control maybe once a week.

Zone 4: A semi-wild area. This zone is mainly used for forage and collecting wild food as well as timber production.

Zone 5: A wild area. There is no human intervention in zone 5 apart from the observation of natural ecosystems and cycles.

Layers

Layers are one of the tools used to design functional ecosystems that are both sustainable and of direct benefit to man. A mature ecosystem has a huge number of relationships between its component parts: trees, understory, ground cover, soil, fungi, insects and animals. Because plants grow to different heights, a diverse community of life is able to grow in a relatively small space, as each layer is stacked one on top of another. There are seven recognised layers in permaculture, although some practitioners also include an eighth layer, fungi.

1. The canopy: the tallest trees in the system. Large trees dominate but do not saturate the area, i.e. there exist patches barren of trees.
2. Low tree layer: dwarf fruit trees, citrus trees and other short trees
3. Shrubs: a diverse layer that includes most berry bushes
4. Herbaceous: may be annuals, biennials or perennials; most annuals will fit into this layer
5. Rhizosphere: root crops including potatoes and other edible tubers
6. Soil surface: cover crops to retain soil and lessen erosion, along with green manures to add nutrients and organic matter to the soil, especially nitrogen
7. Vertical layer: climbers or vines, such as runner beans and lima beans (vine varieties)

Common Design Elements

Agroforestry

Agroforestry is an integrated approach of using the interactive benefits from combining trees and shrubs with crops and/or livestock. It combines agricultural and forestry technologies to create more diverse, productive, profitable, healthy and sustainable land-use systems. In agroforestry systems, trees or shrubs are intentionally used within agricultural systems, or non-timber forest products are cultured in forest settings. Knowledge, careful selection of species and good management of trees and crops are needed to optimise the production and positive effects within the system and to minimise negative competitive effects. Forest gardening is a type of agroforestry that is particularly popular with permaculture designers.

Animals

Animals are often incorporated into the site design. Chickens can be used as a method of weed control and also as a producer of eggs, meat and fertilizer. Bill Mollison stated that "animals represent a valid method of storing inedible vegetation as food."

However not all permaculture sites keep animals for meat, eggs or milk. Sometimes animals function as pets or are treated as co-habitats and co-workers of the site, eating foods normally unpalatable to people such as slugs and termites, being an integral part of the pest management by eating some pests, supplying fertilizer through their droppings and controlling some weed species.

Other projects avoid the use of domesticated animals altogether. Vegan permaculture (also known as veganic permaculture, veganiculture or vegaculture) is essentially the same as permaculture except for the addition of Animalcare as a fourth core value alongside Earthcare, Peoplecare and Fairshare. Zalan Glen, a raw vegan, proposes that *vegaculture* should emerge out of permaculture in the same way veganism split from vegetarianism in the 1940s.

Energy

In passive solar building design, windows, walls, and floors are made to collect, store, and distribute solar energy in the form of heat in the winter and reject solar heat in the summer.

Current industrial agricultural systems of food production are not fully renewable. Industrial agriculture uses large amounts of petroleum and natural gas, both to run the equipment, and to supply pesticides and fertilizers. Permaculture is in part an attempt to create a renewable system of food production that relies upon minimal amounts of energy.

Traditional pre-industrial agriculture was labour intensive, industrial agriculture is fossil fuel intensive, and permaculture is design and information intensive and attempts to be petrofree. Partially permaculture is an attempt to work smarter, not harder; and when possible renewable energy designs such as passive solar should be used.

Peak oil is the point in time when the maximum rate of global petroleum extraction is reached, after which the rate of production enters terminal decline. Mollison and Holmgren have both written extensively on the topic, and it is a much discussed concept amongst permaculturists. Applying permaculture principals means using fewer non-renewable sources of energy, and particularly petroleum based forms of energy.

Natural Building

A natural building involves a range of building systems and materials that place major emphasis on sustainability. Ways of achieving sustainability through natural building focus on durability and the use of minimally processed, plentiful or renewable resources, as well as those that, while recycled or salvaged, produce healthy living environments and maintain indoor air quality. Natural building tends to rely on human labour, more than technology.

The basis of natural building is the need to lessen the environmental impact of buildings and other supporting systems, without sacrificing comfort, health or aesthetics. To be more sustainable, natural building uses primarily abundantly available, renewable, reused or recycled materials. In addition to relying on natural building materials, the emphasis on the architectural design is heightened. The orientation of a building, the utilisation of local climate and site conditions, the emphasis on natural ventilation through design, fundamentally lessen operational costs and positively impact the environmental. Building compactly and minimising the ecological footprint is common, as are on-site handling of energy acquisition, on-site water capture, alternate sewage treatment and water reuse.

Rainwater and Greywater

Rainwater harvesting is the accumulating and storing of rainwater for reuse before it reaches the aquifer. It has been used to provide drinking water, water for livestock, water for irrigation, as well as other typical uses. Rainwater collected from the roofs of houses and local institutions can make an important contribution to the availability of drinking water. It can supplement the subsoil water level and increase urban greenery. Water collected from the ground, sometimes from areas which are especially prepared for this purpose, is called stormwater harvesting.

Greywater is wastewater generated from domestic activities such as laundry, dishwashing, and bathing, which can be recycled on-site for uses such as landscape irrigation and constructed wetlands. This wastewater contains no fecal matter (human feces). Greywater differs from water from the toilets which is designated sewage or blackwater to indicate it contains human waste.

Sheet Mulching

In agriculture and gardening, mulch is a protective cover placed over the soil. Any material or combination can be used as mulch, stones, leaves, plastic, cardboard etc, though in Permaculture mulches of organic material are the most common because they perform more functions. These include: absorbing rainfall, reducing evoporation, providing nutrients, increasing organic matter in the soil, feeding and creating habitat for soil organisms, suppressing weed growth and seed germination, moderating diurnal temperature swings, protecting against frost, and reducing erosion. Sheet mulching is an agricultural no-dig gardening technique that attempts to mimic natural processes occurring within forests, sheet mulching mimics the leaf cover that is found on forest floors. When deployed properly and in combination with other Permacultural principles, it can generate healthy, productive and low maintenance ecosystems.

Sheet mulch serves as a "nutrient bank," storing the nutrients contained in organic matter and slowly making these nutrients available to plants. It also improves the soil by attracting and feeding earthworms, and adding humus. Earthworms "till" the soil, and their worm castings are among the best fertilizers and soil conditioners. Sheet mulching can be used to reduce or eliminate undesirable plants by starving them of light, and may be more advantageous than using herbicide or other methods of control.

Trademark and Copyright Claims

There has been contention over who if anyone controls the legal rights to the word "Permaculture", meaning is it trademarked or copyrighted, and if so, who holds the legal rights to the use of the word. For a long time Bill Mollison claimed to have copyrighted the word permaculture, and his books reflected that on the copyright page, saying "The contents of this book and the word PERMACULTURE are copyright." These statements were largely accepted at face-value within the permaculture community. However, copyright law does not protect names, ideas, concepts, systems, or methods of doing something; it only protects the expression or the description of an idea, not the idea itself. Eventually Mollison acknowledged that he was mistaken and that no copyright protection existed for the word "permaculture".

Mollison explained that the word "permaculture" was copyrighted to protect the quality of teaching, particularly with relation to the Permaculture Design Course (PDC), a 72 hour course usually taught over a period of 14 days. The PDC is a formal means of training an individual the ideas and techniques associated with permaculture. Mollison's argument was if the word was copyrighted, then only those who had been trained and shown to have a reasonable level of proficiency would be allowed to teach the PDC. However, some of those who taught the PDC wanted to adjust the curriculum to better reflect the local conditions of where it was being taught. For example, should a course taught in an urban setting such as New York City be unchanged from what is taught in rural Australia? Mollison was adamant that the curriculum should be taught as he had designed it, without being altered.

In 2000 Mollison's US based Permaculture Institute sought a service mark (a form of trademark) for the word permaculture when used in educational services such as conducting classes, seminars, or workshops. The service mark would have allowed Mollison and his two Permaculture Institutes (one in the US and one in Australia) to set enforceable guidelines as to how permaculture could be taught and who could teach it, particularly with relation to the PDC. The service mark failed and was abandoned in 2001. Also in 2001 Mollison applied for trademarks in Australia for the terms "Permaculture Design Course" and "Permaculture Design". These applications where both withdrawn in 2003. In 2009 he sought a trademark for " Permaculture a Designers' Manual" and "Introduction to Permaculture", the names of two of his books. These applications where withdrawn in 2011. There has never been a trademark for the word Permaculture in Australia.

Criticisms

John Robin has criticized permaculture for its potential to spread environmental weeds. Another critique of permaculture is that the view that woods are more highly productive than farmland on the basis of the theory of ecological succession which states that net productivity declines as ecosystems mature is false. Proponents of permaculture respond that this is only false if one includes data from climax vegetation as opposed to maturing forests. Some critics have questioned whether there is enough scientifically tested data to validate many of the claims promoted by permaculture advocates.

Pesticides

Most organic farms largely avoid pesticides as opposed to conventional farms. Some pesticides damage the environment or with direct exposure, human health. Children exposed to pesticides are of special concern. According to the National Academy of Sciences:

> *"A fundamental maxim of pediatric medicine is that children are not 'little adults.' Profound differences exist between children and adults. Infants and children are growing and developing. Their metabolic rates are more rapid than those of adults. There are differences in their ability to activate, detoxify, and excrete xenobiotic compounds. All these differences can affect the toxicity of pesticides in infants and children, and for these reasons the toxicity of pesticides is frequently different in children and adults."*

The five main pesticides used in organic farming are Bt (a bacterial toxin), pyrethrum, rotenone, copper and sulphur. Fewer than 10% of organic vegetable farmers acknowledge using these pesticides regularly; 5.3% of vegetable growers will admit rotenone use; while 1.7% admit pyrethrum use (Lotter 2003:26). Reduction and elimination of chemical pesticide use is technically challenging. Organic pesticides often complement other pest control strategies.

Figure 1: A cropduster spraying pesticide on a field

Ecological concerns primarily focus around pesticide use, as 16% of the world's pesticides are used in the production of cotton.

Runoff is one of the most damaging effects of pesticide use. The USDA Natural Resources Conservation Service tracks the environmental effects of water contamination and concluded, "the Nation's pesticide policies during the last twenty six years have succeeded in reducing overall environmental risk, in spite of slight increases in area planted and weight of pesticides applied. Nevertheless, there are still areas of the country where there is no evidence of progress, and areas where risk levels for protection of drinking water, fish, algae and crustaceans remain high".

Food Quality and Safety

Many studies have examined the relative quality and safety benefits of organic and conventional agricultural techniques. The results are diverse. Some find no significant differences. Others disagree. An example of the "no differences" school stated:

No evidence of a difference in content of nutrients and other substances between organically and conventionally produced crops and livestock products was detected for the majority of nutrients assessed in this review suggesting that organically and conventionally produced crops and livestock products are broadly comparable in their nutrient content... There is no good evidence that increased dietary intake, of the nutrients identified in this review to be present in larger amounts in organically than in conventionally produced crops and livestock products, would be of benefit to individuals consuming a normal varied diet, and it is therefore unlikely that these differences in nutrient content are relevant to consumer health.

However, they also found that statistically significant differences between the composition of organic and conventional food were present for a few substances.

"Organic products stand out as having higher levels of secondary plant compounds and vitamin C". Organic kiwifruit had more antioxidants.

A review of potential health effects analysed eleven articles, concluding, "because of the limited and highly variable data available, and concerns over the reliability of some reported findings, there is currently no evidence of a health benefit from consuming organic compared to conventionally produced foodstuffs. It should be noted that this conclusion relates to the evidence base currently available on the nutrient content of foodstuffs, which contains limitations in the design and in the comparability of studies."

Individual studies have considered a variety of possible impacts, including pesticide residues. Pesticide residues present a second channel for health effects. Comments include, "Organic fruits and vegetables can be expected to contain fewer agrochemical residues than conventionally grown alternatives; yet, "the significance of this difference is questionable" and "It is intuitive to assume that children whose diets consist of organic food items would have a lower probability of neurologic health risks", and pesticide exposure brought an increased risk for ADHD in one study.

Nitrate concentrations may be less, but the health impact of nitrates is debated. Lack of data has limited research on the health effects of natural plant pesticides and bacterial pathogens. Consumption of organic milk was associated with a decrease in risk for eczema, although no comparable benefit was found for organic fruits, vegetables, or meat.

The higher cost of organic food (ranging from 45 to 200%) could inhibit consumption of the recommended 5 servings per day of vegetables and fruits, which improve health and reduce cancer regardless of their source.

In *Dirt: The Erosion of Civilizations*, geomorphologist David Montgomery outlines a coming crisis from soil erosion. Agriculture relies on roughly one metre of topsoil, and that is being depleted ten times faster than it is being replaced. No-till farming, which some claim depends upon pesticides, is one way to minimise erosion. However, a recent study by the USDA's Agricultural Research Service has found that manure applications in tilled organic farming are better at building up the soil than no-till.

Climate Change

Organic agriculture emphasizes closed nutrient cycles, biodiversity, and effective soil management providing the capacity to mitigate and even reverse the effects of climate change. Organic agriculture can decrease fossil fuel emissions and, like any well managed agricultural system, sequesters carbon in the soil. The elimination of synthetic nitrogen in organic systems decreases fossil fuel consumption by 33 percent and carbon sequestration takes CO_2 out of the atmosphere by putting it in the soil in the form of organic matter which is often lost in conventionally managed soils. Carbon sequestration occurs at especially high levels in organic no-till managed soil.

Agriculture has been undervalued and underestimated as a means to combat global climate change. Soil carbon data show that regenerative organic agricultural practices are among the most effective strategies for mitigating CO_2 emissions.

Nutrient Leaching

Excess nutrients in lakes, rivers, and groundwater can cause algal blooms, eutrophication, and subsequent dead zones. In addition, nitrates are harmful to aquatic organisms by themselves. The main contributor to this pollution is nitrate fertilizers whose use is expected to "double or almost triple by 2050". Organically fertilizing fields "significantly [reduces] harmful nitrate leaching" over conventionally fertilized fields: "annual nitrate leaching was 4.4-5.6 times higher in conventional plots than organic plots".

The large dead zone in the Gulf of Mexico is caused in large part by agricultural runoff: a combination of fertilizer and livestock manure. Over half of the nitrogen released into the Gulf comes from agriculture. This increases costs for fishermen, as they must travel far from the coast to find fish.

Nitrogen leaching into the Danube River was substantially lower among organic farms. The resulting externalities could be neutralised by charging 1 euro per kg of released nitrogen.

Agricultural runoff and algae blooms are strongly linked in California.

Biodiversity

A wide range of organisms benefit from organic farming, but it is unclear whether organic methods confer greater benefits than conventional integrated agri-environmental programs. Nearly all non-crop, naturally occurring species observed in comparative farm land practice studies show a preference for organic farming both by abundance and diversity. An average of 30% more species inhabit organic farms. Birds, butterflies, soil microbes, beetles, earthworms, spiders, vegetation, and mammals are particularly affected. Lack of herbicides and pesticides improve biodiversity fitness and population density. Many weed species attract beneficial insects that improve soil qualities and forage on weed pests. Soil-bound organisms often benefit because of increased bacteria populations due to natural fertilizer such as manure, while experiencing reduced intake of herbicides and pesticides. Increased biodiversity, especially from beneficial soil microbes and mycorrhizae have been proposed as an explanation for the high yields experienced by some organic plots, especially in light of the differences seen in a 21-year comparison of organic and control fields.

Biodiversity from organic farming provides capital to humans. Species found in organic farms enhance sustainability by reducing human input (e.g., fertilizers, pesticides). Farmers that produce with organic methods reduce risk of poor yields by promoting biodiversity. Common game birds such as the ring-necked pheasant and the northern bobwhite often reside in agriculture landscapes, and benefit recreational hunters. The effect of organic farming has been a subject of interest for researchers. Theory suggests that organic farming practices, which exclude the use of synthetic pesticides and fertilizers, would be beneficial for biodiversity, and this has proven true. Spanning all associated species, there is an average of 30% more on organic farms versus conventional farming methods.

In ancient farming practices, farmers did not possess the technology or manpower to have a significant impact on the destruction of biodiversity even as mass-production agriculture was rising. Nowadays, common farming methods generally rely on pesticides to maintain high yields. With such, most agricultural landscapes favour monoculture crops with very little flora or fauna co-existence (van Elsen 2000). Modern organic farm practices such as the removal of pesticides and the inclusion of animal manure, crop rotation, and multi-cultural crops provides the chance for biodiversity to thrive.

Benefits to Biodiversity

Nearly all non-crop, naturally-occurring species observed in comparative farm land practice studies show a preference in organic farming both by population and richness. Spanning all associated species, there is an average of 30% more on organic farms versus conventional farming methods. Birds, butterflies, soil microbes, beetles, earthworms, spiders, vegetation, and mammals are particularly effected. Organic crops use little or no herbicides and pesticides and thus biodiversity fitness and population density benefit. Many weed species attract beneficial insects that improve soil qualities and forage on weed pests. Soil-bound organisms often benefit because of increased bacteria populations due to natural fertilizer spread such as manure, while experiencing reduced intake of herbicides and pesticides commonly associated with conventional farming methods. Increased biodiversity, especially from soil microbes such as mycorhizzae, have been proposed as an explanation for the high yields experienced by some organic plots, especially in light of the differences seen in a 21-year comparison of organic and control fields.

Impact of Increased Biodiversity

The level of biodiversity that can be yielded from organic farming provides a natural capital to humans. Species found in most organic farms provides a means of agricultural sustainability by reducing amount of human input (e.g. fertilizers, pesticides). Farmers that produce with organic methods reduce risk of poor yields by promoting biodiversity. Common game birds such as the ring-necked pheasant and the northern bobwhite often reside in agriculture landscapes, and are a natural capital yielded from high demands of recreational hunting. Because bird species richness and population are typically higher on organic farm systems, promoting biodiversity can be seen as logical and economical.

Highly-impacted Animal Species

Earthworms

Earthworm population and diversity appears to have the most significant data out of all studies. Out of six studies comparing earthworm biodiversity to organic and conventional farming methods, all six suggested a preference for organic practices including a study at the pioneering Haughley farm in 1980/1981 that compared earthworm populations and soil properties after 40 years. Hole et al. (2005) summarised a study conducted by Brown (1999) and found nearly double the population and diversity when comparing farming methods.

Birds

Organic farms are said to be beneficial to birds while remaining economical. Bird species are one of the most prominent animal groups that benefit from organic farming methods. Many species rely on farmland for foraging, feeding, and migration phases. With such, bird populations often relate directly to the natural quality of farmland. The more natural diversity of organic farms provides better habitats to bird species, and is especially beneficial when the farmland is located within a migration zone. In 5 recent studies almost all bird species including locally declining species, both population and variation increased on organic farmland,. Making a switch from conventional farming methods to organic practices also seems to directly improve bird species in the area. While organic farming improves bird populations and diversity, species populations receive the largest boost when organic groups are varied within a landscape.

Butterflies

A specific study done in the UK in 2006 found substantially more butterflies on organic farms versus standard farming methods except for two pest species. The study also observed higher populations in uncropped field margins compared with cropland edges regardless of farm practice. Conversely, Weibull et al. (2000) found no significant differences in species diversity or population.

Spiders

Ten studies have been conducted involving spider species and abundance on farm systems. All but three of the studies indicated that there was a higher diversity of spider species on organic farms, in addition to populations of species. Two of the studies indicated higher species diversity, but statistically insignificant populations between organic and standard farming methods.

Soil Microbes

Out of 13 studies comparing bacteria and fungus communities between organic and standard farming, 8 of the studies showed heightened level of growth on organic farm systems. One study concluded that the use of "green" fertilizers and manures was the primary cause of higher bacterial levels on organic farms. On the other hand, nematode population/diversity depended on what their primary food intake was. Bacteria-feeding nematodes showed preference towards organic systems whereas fungus-feeding nematodes showed preference for standard farm systems. The heightened level of bacteria-feeding nematodes makes sense due to higher levels of bacteria in organic soils, but the fungus-feeding populations being higher on standard farms seems to contradict the data since more fungi are generally found on organic farms.

Beetles

According to Hole et al. (2005), beetle species are among the most commonly studied animal species on farming systems. Twelve studies have found a higher population and species richness of carabids on organic systems. The overall conclusion of significantly higher carabid population species and diversity is that organic farms have a higher level of weed species where they can thrive. Staphylinid populations and diversity have seemed to show no specific preference with some studies showing higher population and diversity, some with lower population and diversity, and one study showed no statistical significance between the organic and conventional farming systems.

Mammals

Two comparative studies have been conducted involving mammal populations and diversity among farm practices. A study done by Brown (1999) found that small mammal population density and diversity did not depend on farming practices, however overall activity was higher on organic farms. It was concluded that more food resources were available to small mammals on organic farms because of the reduction or lack of herbicides and pesticides. Another study conducted by Wickramasinghe et al. (2003) compared bat species and activity. Species activity and foraging were both more than double on organic farms compared to conventional farms. Species richness was also higher on organic farms, and 2 of the sixteen species sighted were found only on organic farms.

Vegetation

Approximately ten studies have been conducted to compare non-crop vegetation between organic and conventional farming practices. Hedgerow, inner-crop and grassland observations were made within these studies and all but one showed a higher weed preference and diversity in or around organic farms. Most of these studies showed significant overall preference for organic farming preferences especially for broad-leafed species, but many grass species showed far less differentiation. It was concluded that broad-leafed species showed far less tolerance to herbicides applied to most conventional farms. Both organic and conventional farms had higher population and richness on farm edges than they did in the middle of croplands. In addition, the weed existence was more obvious on conventional farms likely because pesticide interaction was low or non-existent. Organic farm weed population and richness was believed to be lower in mid-crop land because of weed-removal methods such as under sowing. Switching from conventional to organic farming often results in a "boom" of weed speciation due to intense chemical change of soil composition from the lack of herbicides and pesticides. Natural plant species can also vary on organic farms from year-to-year because crop rotation creates new competition based on the chemical needs of each crop.

Farmers' Benefits from Increased Biodiversity

Biological research on soil and soil organisms has proven beneficial to the system of organic farming. Varieties of bacteria and fungi break down chemicals, plant matter and animal waste into productive soil nutrients. In turn, the producer benefits by healthier yields and more arable soil for future crops. Furthermore, a 21-year study was conducted testing the effects of organic soil matter and its relationship to soil quality and yield. Controls included actively managed soil with varying levels of manure, compared to a plot with no manure input. After the study commenced, there was significantly lower yields on the control plot when compared to the fields with manure. The concluded reason was an increased soil microbe community in the manure fields, providing a healthier, more arable soil system.

Detriments to Biodiversity through Organic Farming

Organic farming practices still require active participation from the farmer to effectively boost biodiversity. Making a switch to organic farming methods does not automatically or guarantee improved biodiversity. Pro-conservation ethics are required to create arable farm land that generates biodiversity. Conservationist ideals are commonly overlooked because they require additional physical and economical efforts from the producer. Common weed-removal processes like undercutting and controlled burning provides little opportunity for species survival, and often leads to comparable populations and richness to conventionally-managed landscapes when performed in excess. Another common process is the addition of biotopes in the form of hedgerows and ponds to further improve species richness. Farmers commonly make the mistake of over-using these resources for more intense crop production because organic yields are typically lower. Another error comes from the over-stratification of biotopes. A series of small clusters does not provide adequate land area for high biodiversity potential.

Sales and Marketing

Most sales are concentrated in developed nations. These products are what economists call credence goods in that they rely on uncertain certification. Interest in organic products dropped between 2006 and 2008, and 42% of Americans polled don't trust organic produce. 69% of Americans claim to occasionally buy organic products, down from 73% in 2005. One theory was that consumers were substituting "local" produce for "organic" produce.

Distributors

In the United States, 75% of organic farms are smaller than 2.5 hectares. In California 2% of the farms account for over half of sales. Small farms join together in cooperatives such as Organic Valley, Inc. to market their goods more effectively.

Most small cooperative distributors have merged or were acquired by large multinationals such as General Mills, Heinz, ConAgra, Kellogg, and others. In 1982 there were 28 consumer cooperative distributors, but as of 2007 only 3 remained. This consolidation has raised concerns among consumers and journalists of potential fraud and degradation in standards. Most sell their organic products through subsidiaries, under other labels.

Organic foods also can be a niche in developing nations. It would provide more money and a better opportunity to compete internationally with the huge distributors. Organic prices are much more stable than conventional foods, and the small farms can still compete and have similar prices with the much larger farms that usually take all of the profits.

Farmers' Markets

Price premiums are important for the profitability of small organic farmers. Farmers selling directly to consumers at farmers' markets have continued to achieve these higher returns. In the United States the number of farmers' markets tripled from 1,755 in 1994 to 5,274 in 2009.

Capacity Building

Organic agriculture can contribute to ecologically sustainable, socio-economic development, especially in poorer countries. The application of organic principles enables employment of local resources (e.g., local seed varieties, manure, etc.) and therefore cost-effectiveness. Local and international markets for organic products show tremendous growth prospects and offer creative producers and exporters excellent opportunities to improve their income and living conditions.

Organic agriculture is knowledge intensive. Globally, capacity building efforts are underway, including localised training material, to limited effect. As of 2007, the International Federation of Organic Agriculture Movements hosted more than 170 free manuals and 75 training opportunities.

Controversy

Norman Borlaug (father of the "Green Revolution" and a Nobel Peace Prize laureate), Prof A. Trewavas and other critics contested the notion that organic agricultural systems are more friendly to the environment and more sustainable than conventional farming systems. Borlaug asserts that organic farming practices can at most feed 4 billion people, after expanding cropland dramatically and destroying ecosystems in the process. The Danish Environmental Protection Agency estimated that phasing out all pesticides would result in an overall yield reduction of about 25%. Environmental and health effects were assumed but hard to assess.

In contrast, the UN Environmental Programme concluded that organic methods greatly increase yields in Africa. A review of over two hundred crop comparisons argued that organic farming could produce enough food to sustain the current human population and that the difference in yields between organic and non-organic methods were small, with non-organic methods yielding slightly more in developed areas and organic methods yielding slightly more in developing areas.

That analysis has been criticised by Alex Avery of the Hudson Institute, who contends that the review claimed many non-organic studies to be organic, misreported organic yields, made false comparisons between yields of organic and non-organic studies which were not comparable, counted high organic yields several times by citing different papers which referenced the same data, and gave equal weight to studies from sources which were not impartial.

The Center for Disease Control repudiated a claim by Avery's father, Dennis Avery (also at Hudson) that the risk of E. coli infection was eight times higher when eating organic food. (Avery had cited CDC as a source.) Avery had included problems stemming from non-organic unpasteurized juice in his calculations. Epidemiologists traced the 2011 E. coli O104:H4 outbreak - which caused over 3,900 cases and 52 deaths - to an organic farm in Bienenbüttel in Germany.

Urs Niggli, director of the FiBL Institute, contends that a global campaign against organic farming derives mostly from Alex Avery's book *The truth about organic farming.*

Organic Farming by Country

Organic farming is practised across the globe, but the markets for sale are strongest in North America and Europe.

The following information is taken from the 2009 edition of the yearbook "The World of Organic Agriculture", published by the International Federation of Organic Movements IFOAM, the Research Institute of Organic Agriculture FiBL and the International Trade Centre ITC.

According to the latest survey on organic agriculture, carried out by the Research Institute of Organic Agriculture FiBL and the International Federation of Organic Agriculture Movements IFOAM, organic agriculture is developing rapidly, and statistical information is now available from 141 countries of the world. Its share of agricultural land and farms continues to grow in many countries. The main results of the global survey on certified organic farming show that 32.2 million hectares of agricultural land are managed organically by more than 1.2 million producers, including smallholders (2007). In addition to the agricultural land, there are 0.4 million hectares of certified organic aquaculture. Global demand for organic products remains robust, with sales increasing by over five billion US Dollars a year. Organic Monitor estimates international sales to have reached 46.1 billion US Dollars in 2007 (FiBL, IFOAM, ITC 2009).

Africa

In Africa, there are almost 900'000 hectares of certified organic agricultural land (end 2007). This constitutes about three percent of the world's organic agricultural land. The countries with the most organic land are Uganda (296'203), Tunisia (154'793 Hectares), and Ethiopia (140'308 hectares). The highest shares of organic land are in Sao Tome and Prince (5 percent), Uganda (2.3 percent) and Tunisia (1.6 percent). The majority of certified organic produce is destined for export markets, with the large majority being exported to the European Union, which is Africa's largest market for agricultural produce. The African market for organic products is still small. Three countries have an organic regulation and seven are in the process of drafting one. A first African Organic Conference will be held in Kampala, Uganda, from May 19–22, 2009. (FiBL, IFOAM, ITC 2009).

Asia

The total organic area in Asia is nearly 2.9 million hectares. This constitutes nine percent of the world's organic agricultural land. 230'000 producers were reported. The leading countries are China (1.6 million hectares) and India (1 million hectares). The highest shares of organic land of all agricultural land are in Timor Leste (seven percent). Organic wild collection areas play a major role in India and China. Production of final processed products is growing, although a majority of production is still fresh produce and field crops with low value-added processing, such as dry or processed raw ingredients. Aquaculture (shrimp and fish) on the other hand, is emerging in China, Indonesia, Vietnam, Thailand, Malaysia and Myanmar. Textiles is another important trend. Sector growth is now also driven by imports, and local markets have taken off in many of the big cities in the South and Eastern part of region besides Japan, South Korea, Taiwan and Singapore. Kuala Lumpur, Manila, Bangkok, Beijing, Shanghai, Jakarta, Delhi, Bangalore and other cities are increasing internal consumption of organic products. Nine organic regulations are in place. In seven countries work on national standards and regulations is in progress.

Europe

As of the end of 2007, 7.8 million hectares in Europe were managed organically by more than 200'000 farms. In the European Union, 7.2 million hectares were under organic management, with more than 180'000 organic farms. 1.9 percent of the European agricultural area and four percent of the agricultural area in the European Union is organic. Twenty-four percent of the world's organic land is in Europe. The countries with the largest organic area are Italy (1'150'253 hectares), Spain (988'323 hectares) and Germany (865'336 hectares). The highest percentages are in Liechtenstein (29 percent), Austria (13 percent) and Switzerland (11 percent). Compared to 2006, organic land increased by more than 0.3 million hectares. Sales of organic products were approximately 16 billion Euros in 2007. The largest market for organic products in 2007 was Germany with a turnover of 5.3 billion Euros (2008: 5.8 billion Euros), followed by the UK (2.6 billion Euros), France and Italy (both 1.9 billion Euros). As a portion of the total market share, the highest levels have been reached in Austria, Denmark and Switzerland, with around five percent for organic products. The highest per capita spending is also in these countries. Support for organic farming in the European Union and the neighbouring countries includes grants under rural development programs, legal protection and a European as well as national action plans. One of the key instruments of the European Action Plan on organic food and farming, an information campaign, was launched during 2008, with the aim of increasing awareness of organic farming throughout the European Union. Furthermore, most EU member states have national action plans. In order to boost organic farming research, a technology platform joining the efforts of industry and civil society in defining organic research priorities and defending them vis-à-vis the policy-makers was launched in December 2008. The platform's vision paper reveals the potential of organic food production to mitigate some of the major global problems from climate change and food security, to the whole range of socio-economic challenges in the rural areas).

North America

In North America, almost 2.2 million hectares are managed organically, representing approximately a 0.6 percent share of the total agricultural area. Currently, the number of farms is 12'064. The major part of the organic land is in the US (1.6 million hectares in 2005). Seven percent of the world's organic agricultural land is in North America. Valued at more than 20 billion US Dollars in 2007 (Organic Monitor), the North American market accounted for 45 percent of global revenues. Growing consumer demand for healthy & nutritious foods and increasing distribution in conventional grocery channels are the major drivers of market growth.

The U.S. organic industry grew 21 percent in sales in 2006, and was forecast to experience 18 percent sales growth each year on average from 2007 through 2010. Whether this rate will actually be realised is uncertain due to the economic downturn and reduction in consumer spending in the last quarter of 2008.

Likewise, a downturn is expected in Canada, even though the market growth in Canada, paired with the introduction of the new organic regulations, should provide a good outlook over the coming years. In the United States, the National Organic Program has been in force since 2002. Canada has had a strong organic standard since 1999; this had been, however, voluntary and not supported by regulation. Canada's Organic Product Regulation will be fully implemented on June 30, 2009. Canadian labeling requirements will very similar to those of the US and the EU. In 2008, the new Farm Bill was passed by the US Congress. Increasing expenditures on organic agriculture and programs to approximately 112 million US Dollars1 over the course of its five-year life, the 2008 Farm Bill provides a fivefold increase for the organic sector compared with federal funding in the previous bill.(FiBL, IFOAM, ITC 2009).

Latin America and the Caribbean

In Latin America, 220'000 producers managed 6.4 million hectares of agricultural land organically in 2007. This constitutes 20 percent of the world's organic land. The leading countries are Argentina (2'777'959 hectares), Brazil (1'765'793 hectares) and Uruguay (930'965 hectares). The highest shares of organic agricultural land are in the Dominican Republic and Uruguay with more than six percent and in Mexico and Argentina with more than two percent. Most organic production in Latin America is for export. Important crops are tropical fruits, grains and cereals, coffee and cocoa, sugar and meats. Most organic food sales in the domestic markets of the countries occurs in major cities, such as Buenos Aires and São Paulo.

Fifteen countries have legislation on organic farming, and four additional countries are currently developing organic regulations.

Costa Rica and Argentina have both attained third country status according to the EU regulation on organic farming.

In recognition of the growing importance of the organic sector to Latin America's agricultural economy, governmental institutions have begun to take steps towards increasing involvement; governments are beginning to play a central role in the promotion of organic agriculture. The types of support in Latin American countries range from organic agriculture promotion programs to market access support by export agencies. In a few countries, limited financial support is being given to pay certification cost during the conversion period.

An important process underway in many Latin America countries is the establishment of regulations and standards for the organic sector (FiBL, IFOAM, ITC 2009).

Oceania

This region includes Australia, New Zealand, and island states like Fiji, Papua New Guinea, Tonga and Vanuatu. Altogether, there are 7'222 producers, managing almost 12.1 million hectares. This constitutes 2.6 percent of the agricultural land in the area and 38 percent of the world's organic land. Ninety-nine percent of the organically managed land in the region is in Australia (12 million hectares, 97 percent extensive grazing land), followed by New Zealand (65'000 hectares) and Vanuatu (8'996 hectares). The highest shares of all agricultural land are in Vanuatu (6.1 percent), Samoa (5.5 percent) and the Solomon Islands (3.1 percent). Growth in the organic industry in Australia, New Zealand and the Pacific Islands has been strongly influenced by rapidly growing overseas demand; domestic markets are, however, growing. In New Zealand, a key issue is lack of production to meet growing demand.

Australia has had national standards for organic and biodynamic products in place since 1992, and like New Zealand, it is on the third country list of the European Union. It is expected that the Australian Standard, based on the National Standard employed since the early 1990s for the export market, will be adopted in 2009. In New Zealand, a National Organic Standard was launched in 2003. There is little government support to encourage organic agriculture in Australia. However, over the recent past, governments have been supportive of the Australian Standards issue. Furthermore, funding is made available to promote an understanding among consumers. In New Zealand, through the establishment of the sector umbrella organisation Organics Aotearoa New Zealand and the Organic Advisory Programme as well as other initiatives, there is political recognition of the benefits of organic agriculture.

In the Pacific Islands work on a regional strategy and national plans to lay the foundation of sustainable organic agriculture development in the region is in progress. The Regional Organic Task Force, a technical group representing all sectors and countries involved in organics, was charged with developing the Pacific Standard and will be responsible for implementing the Regional Action Plan. Pacific High Level Organics Group consists of Pacific leaders who have shown a commitment to the development of organic agriculture in the region and provide high level political support and advocacy. The first Pacific Organic Standard was endorsed by Pacific Leaders in September 2008. This provides a platform for further regional policy development around organic agriculture.

Chapter 3

Agroecology

Agroecology is the application of ecological principles to the production of food, fuel, fibre, and pharmaceuticals. The term encompasses a broad range of approaches, and is considered "a science, a movement, [and] a practice."

The prefix *agro-* refers to *agriculture*.

Ecological Strategy

Agroecologists study a variety of agroecosystems, and the field of agroecology is not associated with any one particular method of farming, whether it be organic, conventional, intensive or extensive. Furthermore, it is not defined by certain management practices, such as the use of natural enemies in place of insecticides, or polyculture in place of monoculture.

Additionally, agroecologists do not unanimously oppose technology or inputs in agriculture but instead assess how, when, and if technology can be used in conjunction with natural, social and human assets. Agroecology proposes a context- or site-specific manner of studying agroecosystems, and as such, it recognises that there is no universal formula or recipe for the success and maximum well-being of an agroecosystem. Instead, agroecologists may study questions related to the four system properties of agroecosystems: productivity, stability, sustainability and equitability. As opposed to disciplines that are concerned with only one or some of these properties, agroecologists see all four properties as interconnected and integral to the success of an agroecosystem. Recognising that these properties are found on varying spatial scales, agroecologists do not limit themselves to the study of agroecosystems at any one scale: farm, community, or global.

Agroecologists study these four properties through an interdisciplinary lens, using natural sciences to understand elements of agroecosystems such as soil properties and plant-insect interactions, as well as using social sciences to understand the effects of farming practices on rural communities, economic constraints to developing new production methods, or cultural factors determining farming practices.

Various Approaches

Agroecologists do not always agree about what agroecology is or should be in the long-term. Different definitions of the term agroecology can be distinguished largely by the specificity with which one defines the term "ecology," as well as the term's potential political connotations. Definitions of agroecology, therefore, may be first grouped according to the specific contexts within which they situate agriculture.

Agroecology is defined by the OECD as "the study of the relation of agricultural crops and environment." This definition refers to the "-ecology" part of "agroecology" narrowly as the natural environment. Following this definition, an agroecologist would study agriculture's various relationships with soil health, water quality, air quality, meso- and micro-fauna, surrounding flora, environmental toxins, and other environmental contexts.

A more common definition of the word can be taken from Dalgaard et al., who refer to agroecology as the study of the interactions between plants, animals, humans and the environment within agricultural systems. Consequently, agroecology is inherently multidisciplinary, including factors from agronomy, ecology, sociology and economics. In this case, the "-ecology" portion of "agroecology is defined broadly to include social, cultural, and economic contexts as well.

Agroecology is also defined differently according to geographic location. In the global south, the term often carries overtly political connotations. Such political definitions of the term usually ascribe to it the goals of social and economic justice; special attention, in this case, is often paid to the traditional farming knowledge of indigenous populations. North American and European uses of the term sometimes avoid the inclusion of such overtly political goals. In these cases, agroecology is seen more strictly as a scientific discipline with less specific social goals. Fred Buttel makes a more academic distinction of the various approaches within the field, separating it into five broad categories:

Ecosystems Agroecology

This approach is driven by the ecosystems biology of Eugene Odum. This approach is based in the hypotheses that the natural systems, with its stability and resilience, provide the best model to mimic if sustainability is the goal. Normally, ecosystems agroecology is not actively involved in social science; however, this school is essentially based on the belief that large-scale agriculture is inappropriate. The work of Steve Gliessman is prototypical of this approach.

Agronomic Ecology

The basic approach in this branch is derived mostly from agronomy, including the traditional agricultural production sciences. This approach also does not actively involve social sciences in the agroecological analysis, but uses social sciences to understand the processes by which agriculture became unsustainable. Chuck Francis, Richard Hardwood, Ricardo Salvador, and Matt Liebman are exemplars of this approach.

Ecological Political Economy

The driving force behind this form of agroecology is a political-economical critique of modern agriculture. The school believes that only radical changes in political economy and the moral economy of research will reduce the negative costs of modern agriculture. The works of Miguel Altieri (ecosystem biologist), John Vandermeer (population ecologist), Richard Lewontin, and Richard Levins provide examples of this politically charged and socially-oriented version of agroecology.

Agro-population Ecology

This approach is derived from the science of ecology primarily based on population ecology, which over the past three decades has been displacing the ecosystems biology of Odum. Buttel explains the main difference between the two categories, saying that "the application of population ecology to agroecology involves the primacy not only of analysing agroecosystems from the perspective of the population dynamics of their constituent species, and their relationships to climate and biogeochemistry, but also there is a major emphasis placed on the role of genetics." David Andow and Alison Power are cited as examples of professionals espousing this view.

Integrated Assessment of Multifunctional Agricultural Systems

This approach focuses on the multifunctionality of the landscape, instead of focusing solely on the agricultural enterprise. Agriculture and the food system are considered parts of an institutional complex that relates to and integrates with other social institutions. Scholars adopting this highly integrated approach, mostly Europeans, do not consider any one discipline the leader of agroecology.

Holon Agroecology

First introduced in 2007 by the soil scientist William L. Bland and the environmental sociologist Michael M. Bell of the University of Wisconsin–Madison, holon agroecology draws on Koestler's notion of a "holon" which is both part and whole and develops it with ideas of narrative, intentionality, and incompleteness or unfinalizability, within an ever-changing "ecology of contexts". In contrast to systems thinking, holon agroecology stresses seeing the agricultural endeavour as an unfinished accomplishment that is constantly adjusting itself to its many contexts and their conflicts and incommensurabilities. The farm holon represents a kind of "holding together" in order to persist through change, but a holding together that is never fully unified and worked out.

History

Pre-WWII

The notions and ideas relating to crop ecology have been around since at least 1911 when F.H. King released *Farmers of Forty Centuries*. King was one of the pioneers as a proponent of more quantitative methods for characterisation of water relations and physical properties of soils. In the late 1920s the attempt to merge agronomy and ecology was born with the development of the field of crop ecology. Crop ecology's main concern was where crops would be best grown. Actually, it was only in 1928 that agronomy and ecology were formally linked by Klages.

The first mention of the term agroecology was in 1928, with the publication of the term by Bensin in 1928. The book of Tischler (1965), was probably the first to be actually titled 'agroecology'. He analysed the different components (plants, animals, soils and climate) and their interactions within an agroecosystem as well as the impact of human agricultural management on these components.

Other books dealing with agroecology, but without using the term explicitly were published by the German zoologist Friederichs (1930) with his book on agricultural zoology and related ecological/ environmental factors for plant protection and by American crop physiologist Hansen in 1939 when both used the word as a synonym for the application of ecology within agriculture.

Post-WWII

Gliessman mentions that post-WWII, groups of scientists with ecologists gave more focus to experiments in the natural environment, while agronomists dedicated their attention to the cultivated systems in agriculture.

According to Gliessman, the two groups kept their research and interest apart until books and articles using the concept of agroecosystems and the word agroecology started to appear in 1970. Dalgaard explains the different points of view in ecology schools, and the fundamental differences, which set the basis for the development of agroecology. The early ecology school of Henry Gleason investigated plant populations focusing in the hierarchical levels of the organism under study.

Friederich Clement's ecology school, however included the organism in question as well as the higher hierarchical levels in its investigations, a "landscape perspective". However, the ecological schools where the roots of agroecology lie are even broader in nature. The ecology school of Tansley, whose view included both the biotic organism and their environment, is the one from which the concept of agroecosystems emerged in 1974 with Harper.

In the 1960s and 70's the increasing awareness of how humans manage the landscape and its consequences set the stage for the necessary cross between agronomy and ecology. Even though, in many ways the environmental movement in the US was a product of the times, the Green Decade spread an environmental awareness of the unintended consequences of changing ecological processes. Works such as *Silent Spring*, and *The Limits to Growth*, and changes in legislation such as the Clean Air Act, Clean Water Act, and the National Environmental Policy Act caused the public to be aware of societal growth patterns, agricultural production, and the overall capacity of the system.

Fusion with Ecology

After the 1970s, when agronomists saw the value of ecology and ecologists began to use the agricultural systems as study plots, studies in agroecology grew more rapidly. Gliessman describes that the innovative work of Prof. Efraim Hernandez X., who developed research based on indigenous systems of knowledge in Mexico, led to education programs in agroecology. In 1977 Prof. Efraim Hernandez X. explained that modern agricultural systems had lost their ecological foundation when socio-economic factors became the only driving force in the food system. The acknowledgement that the socio-economic interactions are indeed one of the fundamental components of any agroecosystems came to light in 1982, with the article Agroecologia del Tropico Americano by Montaldo. The author argues that the socio-economic context cannot be separated from the agricultural systems when designing agricultural practices.

In 1995 Edens et al. in Sustainable Agriculture and Integrated Farming Systems solidified this idea proving his point by devoting special sections to economics of the systems, ecological impacts, and ethics and values in agriculture. Actually, 1985 ended up being a fertile and creative year for the new discipline.

For instance in the same year, Miguel Altieri integrated how consolidation of the farms, and cropping systems impact pest populations. In addition, Gliessman highlighted that socio-economic, technological, and ecological components give rise to producer choices of food production systems. These pioneering agroecologists have helped to frame the foundation of what we today consider the interdisciplinary field of agroecology.

Publications

Year	*Author(s)*	*Title*
1928	Klages	Crop ecology and ecological crop geography in the agronomic curriculum
1939	Hanson	Ecology in agriculture
1956	Azzi	Agricultural ecology
1965	Tischler	Agrarökologie
1973	Janzen	Tropical agroecosystems
1974	Harper	The need for a focus on agro-ecosystems
1976	Loucks	Emergence of research on agroecosystems
1977	Hernanez Xolocotzi	Agroecosistemas de Mexico
1978	Gliessman	Agroecosistemas y tecnologia agricola tradicional
1979	Hart	Agroecosistemas: conceptos básicos
1979	Cox & Atkins	Agricultural ecology: an analysis of world food production systems
1980	Hart	Agroecosistemas

Contd...

Year	Author(s)	Title
1981	Gliessman, Garcia & Amador	The ecological basis for the application of traditional agricultural technology in the management of tropical agroecosystems
1982	Montaldo	Agroecologia del trópico americano
1983	Altieri	Agroecology
1984	Lowrance, Stinner & House	Agricultural ecosystems: unifying concepts
1985	Conway	Agroecosystems analysis
1987	Altieri	Agroecology: the scientific basis of alternative agriculture
1990	Allen, Dusen, Lundy, & Gliessman	Integrating social, environmental, and economic issues in sustainable agriculture
1990	Gliessman	Agroecology: researching the ecological basis for sustainable agriculture
1990	Carroll, Vandermeer & Rosset	Agroecology
1990	Altieri & Hecht	Agroecology and small farm development
1991	Caporali	Ecologia per l'agricultura
1991	Bawden	Systems thinking in agriculture
1993	Coscia	Agricultura sostenible
1998	Gliessman	Agroecology: ecological processes in sustainable agriculture
2001	Flora	Interactions between agroecosystems and rural communities
2001	Gliessman	Agroecosystem sustainability
2002	Dalgaard, Porter & Hutchings	Agroecology, scaling, and interdisciplinarity
2003	Francis et al.	Agroecology: The Ecology of Food Systems
2004	Clements, Shrestra	New Dimension in Agroecology
2007	Bland and Bell	A Holon Approach to Agroecology
2007	Gliessman	Agroecology: The Ecology of Sustainable Food Systems

Contd...

Year	*Author(s)*	*Title*
2007	Warner	Agroecology in Action
2009	Wezel, Soldat	A quantitative and qualitative historical analysis of the scientific discipline agroecology
2009	Wezel et al.	Agroecology as a science, a movement or a practice. A review

Applications

To emit a point of view about a particular way of farming, an agroecologist would first seek to understand the contexts in which the farm(s) is(are) involved. Each farm may be inserted in a unique combination of factors or contexts. Each farmer may have their own premises about the meanings of an agricultural endeavour, and these meanings might be different than those of agroecologists. Generally, farmers seek a configuration that is viable in multiple contexts, such as family, financial, technical, political, logistical, market, environmental, spiritual. Agroecologists want to understand the behaviour of those who seek livelihoods from plant and animal increase, acknowledging the organisation and planning that is required to run a farm.

Views on Organic and Non-organic Milk Production

Because organic agriculture proclaims to sustain the health of soils, ecosystems, and people, it has much in common with Agroecology; this does not mean that Agroecology is synonymous with organic agriculture, nor that Agroecology views organic farming as the 'right' way of farming. Also, it is important to point out that there are large differences in organic standards among countries and certifying agencies.

Three of the main areas that agroecologists would look at in farms, would be: the environmental impacts, animal welfare issues, and the social aspects.

Environmental impacts caused by organic and non-organic milk production can vary significantly. For both cases, there are positive and negative environmental consequences.

Compared to conventional milk production, organic milk production tends to have lower eutrophication potential per ton of milk or per hectare of farmland, because it potentially reduces leaching of nitrates (NO_3^-) and phosphates (PO_4^-) due to lower fertilizer application rates. Because organic milk production reduces pesticides utilisation, it increases land use per ton of milk due to decreased crop yields per hectare. Mainly due to the lower level of concentrates given to cows in organic herds, organic dairy farms generally produce less milk per cow than conventional dairy farms. Because of the increased use of roughage and the, on-average, lower milk production level per cow, some research has connected organic milk production with increases in the emission of methane.

Animal welfare issues vary among dairy farms and are not necessarily related to the way of producing milk (organically or conventionally).

A key component of animal welfare is freedom to perform their innate (natural) behaviour, and this is stated in one of the basic principles of organic agriculture. Also, there are other aspects of animal welfare to be considered - such as freedom from hunger, thirst, discomfort, injury, fear, distress, disease and pain. Because organic standards require loose housing systems, adequate bedding, restrictions on the area of slatted floors, a minimum forage proportion in the ruminant diets, and tend to limit stocking densities both on pasture and in housing for dairy cows, they potentially promote good foot and hoof health. Some studies show lower incidence of placenta retention, milk fever, abomasums displacement and other diseases in organic than in conventional dairy herds. However, the level of infections by parasites in organically managed herds is generally higher than in conventional herds.

Social aspects of dairy enterprises include life quality of farmers, of farm labour, of rural and urban communities, and also includes public health.

Both organic and non-organic farms can have good and bad implications for the life quality of all the different people involved in that food chain. Issues like labour conditions, labour hours and labour rights, for instance, do not depend on the organic/non-organic characteristic of the farm; they can be more related to the socio-economical and cultural situations in which the farm is inserted, instead.

As for the public health or food safety concern, organic foods are intended to be healthy, free of contaminations and free from agents that could cause human diseases. Organic milk is meant to have no chemical residues to consumers, and the restrictions on the use of antibiotics and chemicals in organic food production has the purpose to accomplish this goal. But dairy cows in organic farms, as in conventional farms, indeed do get exposed to virus, parasites and bacteria that can contaminate milk and hence humans, so the risks of transmitting diseases are not eliminated just because the production is organic.

In an organic dairy farm, an agroecologist could evaluate the following:

1. Can the farm minimise environmental impacts and increase its level of sustainability, for instance by efficiently increasing the productivity of the animals to minimise waste of feed and of land use?
2. Are there ways to improve the health status of the herd (in the case of organics, by using biological controls, for instance)?
3. Does this way of farming sustain good quality of life for the farmers, their families, rural labour and communities involved?

Views on No-till Farming

No-tillage is one of the components of conservation agriculture practices and is considered more environmental friendly than complete tillage. Due to this belief, it could be expected that agroecologists would not recommend the use of complete tillage and would rather recommend no-till farming, but this is not always the case. In fact, there is a general consensus that no-till can increase soils capacity of acting as a carbon sink, especially when combined with cover crops.

No-till can contribute to higher soil organic matter and organic carbon content in soils, though reports of no-effects of no-tillage in organic matter and organic carbon soil contents also exist, depending on environmental and crop conditions. In addition, no-till can indirectly reduce CO_2 emissions by decreasing the use of fossil fuels.

Most crops can benefit from the practice of no-till, but not all crops are suitable for complete no-till agriculture. Crops that do not perform well when competing with other plants that grow in untilled soil in their early stages can be best grown by using other conservation tillage practices, like a combination of strip-till with no-till areas. Also, crops which harvestable portion grows underground can have better results with strip-tillage, mainly in soils which are hard for plant roots to penetrate into deeper layers to access water and nutrients.

The benefits provided by no-tillage to predators may lead to larger predator populations, which is a good way to control pests (biological control), but also can facilitate predation of the crop itself. In corn crops, for instance, predation by caterpillars can be higher in no-till than in conventional tillage fields.

In places with rigorous winter, untilled soil can take longer to warm and dry in spring, which may delay planting to less ideal dates. Another factor to be considered is that organic residue from the prior year's crops laying on the surface of untilled fields can provide a favourable environment to pathogens, helping to increase the risk of transmitting diseases to the future crop. And because no-till farming provides good environment for pathogens, insects and weeds, it can lead farmers to a more intensive use of chemicals for pest control. Other disadvantages of no-till include underground rot, low soil temperatures and high moisture.

Based on the balance of these factors, and because each farm has different problems, agroecologists will not atest that only no-till or complete tillage is the right way of farming. Yet, these are not the only possible choices regarding soil preparation, since there are intermediate practices such as strip-till, mulch-till and ridge-till, all of them - just as no-till - categorised as conservation tillage. Agroecologists, then, will evaluate the need of different practices for the contexts in which each farm is inserted.

In a no-till system, an agroecologist could ask the following:

1. Can the farm minimise environmental impacts and increase its level of sustainability; for instance by efficiently increasing the productivity of the crops to minimise land use?
2. Does this way of farming sustain good quality of life for the farmers, their families, rural labour and rural communities involved?

By Region

The principles of agroecology are expressed differently depending on local ecological and social contexts.

Latin America

Latin America's experiences with North American Green Revolution agricultural techniques have opened space for agroecologists. Traditional or indigenous knowledge represents a wealth of possibility for agroecologists, including "exchange of wisdoms."

Madagascar

Most of the historical farming in Madagascar has been conducted by indigenous peoples. The French colonial period disturbed a very small percentage of land area, and even included some useful experiments in sustainable forestry. Slash-and-burn techniques, a component of some shifting cultivation systems have been practised by natives in Madagascar for centuries. As of 2006 some of the major agricultural products from slash-and-burn methods are wood, charcoal and grass for Zebu grazing. These practices have taken perhaps the greatest toll on land fertility since the end of French rule, mainly due to overpopulation pressures.

Biodynamic Agriculture

Biodynamic agriculture is a method of organic farming that emphasizes the holistic development and interrelationships of the soil, plants and animals as a self-sustaining system. Biodynamic farming has much in common with other organic approaches, such as emphasizing the use of manures and composts and excluding of the use of artificial chemicals on soil and plants. There are independent certification agencies for biodynamic products; most of these agencies are members of the international biodynamics standards group Demeter International.

Biodynamics was one of the first modern ecological farming systems and is considered to be one of the most sustainable. As of 2011 biodynamic techniques were used on 142,482 hectares in 47 countries; Germany accounts for 45.1% of the global total. Methods unique to the biodynamic approach include the use of fermented herbal and mineral preparations as compost additives and field sprays and the use of an astronomical sowing and planting calendar. Biodynamics originated out of the work of Rudolf Steiner, the founder of anthroposophy.

Biodynamic agriculture has been characterised as pseudoscience.

History

The development of biodynamic agriculture began in 1924 with a series of eight lectures on agriculture given by philosopher Rudolf Steiner at Schloss Koberwitz in Silesia, Germany, (now Kobierzyce in Poland east of Wroc³aw). The course was held in response to a request by farmers who noticed degraded soil conditions and a deterioration in the health and quality of crops and livestock resulting from the use of chemical fertilizers. At the course of eight lectures there were 111 attendees coming from six countries. An agricultural research group was subsequently formed to test the effects of biodynamic methods on the life and health of soil, plants and animals.

Today biodynamics is practiced in more than 50 countries worldwide. Demeter International is the primary certification agency for farms and gardens using the methods.

Geographic Developments

- Steiner's *Agriculture Course* was published in November 1924; an English translation appeared in 1928.
- In Australia the first biodynamic preparations were made by Ernesto Genoni in Melbourne in 1927 and by Bob Williams in Sydney in 1939. Since the 1950s research work has continued at the Biodynamic Research Institute (BDRI) in Powelltown, near Melbourne Australia under the direction of Alex Podolinsky. In 1989 Biodynamic Agriculture Australia was established, as a not for profit association.
- In 1938 Ehrenfried Pfeiffer's *Bio-Dynamic Farming and Gardening* was published in English, German, Dutch, French, and Italian.
- In the United States, the Biodynamic Farming & Gardening Association was founded in 1938 as a New York state corporation.
- In the UK in 1939, Britain's first biodynamic agriculture conference, the Betteshanger Summer School and Conference on Biodynamic Agriculture, was held at Lord Northbourne's farm in Kent; Ehrenfried Pfeiffer was the lead presenter.
- In France the International Federation of Organic Agriculture Movements (IFOAM) was formed in 1972 with five founding members, one of which was the Swedish Biodynamic Association.
- The University of Kassel had a Department of Biodynamic Agriculture from 2006 to March 2011.
- Germany, Italy and India are reported to be the leading countries in biodynamic agriculture based on biodynamic hectares.

Biodynamic Method of Farming

Biodynamic agriculturalists conceive of the farm as an organically self-contained entity with its own individuality, within which organisms are interdependent. "Emphasis is placed on the integration of crops and livestock, recycling of nutrients, maintenance of soil, and the health and well being of crops and animals; the farmer too is part of the whole." Cover crops, green manures and crop rotations are used extensively and the farms foster bio-diversity.

Biodynamic farms often have a cultural component and encourage local community. Some biodynamic farms use the Community Supported Agriculture model, which has connections with social threefolding.

Biodynamic Preparations

Steiner prescribed nine different preparations to aid fertilization which are the cornerstone of biodynamic agriculture, and described how these were to be prepared. Steiner believed that these preparations transferred supernatural terrestrial and cosmic "forces" into the soil. The prepared substances are numbered 500 through 508, where the first two are used for preparing fields whereas the latter seven are used for making compost. A long term trial (DOK experiment) evaluating the biodynamic farming system in comparison with organic and conventional farming systems, found that preparations have influence on soil structure and micro-organisms enhancing soil fertility and increasing biodiversity. Regarding compost development beyond accelerating the initial phase of composting, some positive effects have been noted:

- The field sprays contain substances that stimulate plant growth include cytokinins.
- Some improvement in nutrient content of compost.

Field Preparations

Field preparations, for stimulating humus formation:

- 500: (horn-manure) a humus mixture prepared by filling the horn of a cow with cow manure and burying it in the ground (40–60 cm below the surface) in the autumn. It is left to decompose during the winter and recovered for use the following spring.
- 501: Crushed powdered quartz prepared by stuffing it into a horn of a cow and buried into the ground in spring and taken out in autumn. It can be mixed with 500 but usually prepared on its own (mixture of 1 tablespoon of quartz powder to 250 liters of water) The mixture is sprayed under very low pressure over the crop during the wet season, in an attempt to prevent fungal diseases. It should be sprayed on an overcast day or early in the morning to prevent burning of the leaves.

Both 500 and 501 are used on fields by stirring about one teaspoon of the contents of a horn in 40–60 liters of water for an hour and whirling it in different directions every second minute. Although some biodynamic beliefs refer to buried quartz "fermenting", a 2004 review commented that it is unclear what this actually means, as rock does not ferment.

Compost Preparations

Compost preparations, used for preparing compost, employ herbs which are frequently used in medicinal remedies:

- 502: Yarrow blossoms (*Achillea millefolium*) are stuffed into urinary bladders from Red Deer (*Cervus elaphus*), placed in the sun during summer, buried in earth during winter and retrieved in the spring.
- 503: Chamomile blossoms (*Matricaria recutita*) are stuffed into small intestines from cattle buried in humus-rich earth in the autumn and retrieved in the spring.
- 504: Stinging nettle (*Urtica dioica*) plants in full bloom are stuffed together underground surrounded on all sides by peat for a year.
- 505: Oak bark (*Quercus robur*) is chopped in small pieces, placed inside the skull of a domesticated animal, surrounded by peat and buried in earth in a place where lots of rain water runs past.
- 506: Dandelion flowers (*Taraxacum officinale*) is stuffed into the peritoneum of cattle and buried in earth during winter and retrieved in the spring.
- 507: Valerian flowers (*Valeriana officinalis*) are extracted into water.
- 508: Horsetail (*Equisetum*)

One to three grams (a teaspoon) of each preparation is added to a dung heap by digging 50 cm deep holes with a distance of 2 metres from each other, except for the 507 preparation, which is stirred into 5 liters of water and sprayed over the entire compost surface. All preparations are thus used in homeopathic quantities. Each compost preparation is designed to guide a particular decomposition process in the composting mass.

One study found that the oak bark preparation improved disease resistance in zucchini.

Astronomical Planting Calendar

The approach considers that there are astronomical influences on soil and plant development, specifying, for example, what phase of the moon is most appropriate for planting, cultivating or harvesting various kinds of crops. This aspect of biodynamics has been termed "astrological" in nature.

Seed Production

Biodynamic agriculture has focused on open pollination of seeds (permitting farmers to grow their own seed) and the development of locally adapted varieties. The seed stock is not controlled by large, multinational seed companies.

Trademark Protection of Term Biodynamic

The term *Biodynamic* is a trademark held by the Demeter association of biodynamic farmers for the purpose of maintaining production standards used both in farming and processing foodstuffs. The trademark is intended to protect both the consumer and the producers of biodynamic produce. Demeter International is an organisation of member countries; each country has its own Demeter organisation which is required to meet international production standards (but can also exceed them). The original Demeter organisation was founded in 1928; the U.S. Demeter Association was formed in the 1980s and certified its first farm in 1982. In France, Biodivin certifies biodynamic wine. In Egypt, SEKEM has created the Egyptian Biodynamic Association (EBDA), an association that provides training for farmers to become certified.

Studies of Efficacy

Studies have compared biodynamic farming methods to both other organic methods and to conventional methods. Most studies have found that biodynamic farms have soil quality significantly better than conventionally farmed soils but comparable to the soil quality achieved by other organic methods; the decisive factor is likely to be the use of compost. Studies of yields differ in their conclusions, and have found:

- A long-term study conducted at a commercial vineyard in California compared vineyard blocks treated with biodynamic preparations alongside those tended with general organic farming methods, to examine effects upon soil and crop quality. "No differences were found in soil quality" during the first six years of the study, and analyses of other indicators including the yield per vine, clusters per vine, cluster and berry weight also showed there were no differences. The study did find a statistically significant (p-value < 0.05) difference in the yield-to-pruning weight ratio, indicating an "ideal vine balance for producing high-quality wine grapes" for the biodynamically treated crop, but noted the control vines had been "slightly overcropped". In one particular year of the study the biodynamically treated wine grapes had significantly higher Brix and notably higher total phenols and anthocyanins. In conclusion, the study found that biodynamic preparations "may affect" the vine canopy and chemistry, but showed no effects on the soil and tissue nutrient parameters measured in the study.

- Decomposition was significantly faster in plots which received farmyard manure (FYM) treated with biodynamic preparations than in plots which received no FYM, FYM without preparations or FYM with an alternative preparation. "The application of completely prepared FYM led to significantly higher biomass and abundance of endogeic or anecic earthworms than in plots where non-prepared FYM was applied."
- A 21-year study by Mäder *et al* for the FiBL Institute in Switzerland compared the agronomic and ecological performance of biodynamic, organic and two conventional systems. The study found that nutrient input in the biodynamic and organic systems was 34 to 51% lower than in the conventional systems but crop yield was only 20% lower on average, indicating more efficient production. The total energy (for fuel, production of mineral fertilizer and pesticides, etc.) to produce a dry-matter unit of crop was 20 to 56% lower for the biodynamic and organic systems, and pesticide input was reduced by 97% (by 100% for the biodynamic system). In regard to soil aggregate stability, soil pH, humus formation, soil calcium, microbial biomass, and faunal biomass (earthworms and arthropods), the biodynamic system was superior even to the organic system, which in turn had superior results over the conventional systems. With the significant increase in microbial diversity in the biodynamic and organic systems, there was a significant associated decrease in metabolic quotient, indicating a greater ability to use organic material for plant growth.

The methodology of this study was criticized as faulty by two wine writers: philosophy professor Douglass Smith and criminology professor Jesús Barquín. They said the methodology yielded false conclusions. Smith and Barquín wrote that, to separate the mystical elements in biodynamics from standard organic techniques, no comparison should ever be made between conventional and biodynamic farming, only between biodynamic and organic farming. They noted that the Mäder paper's supplemental materials, found but not in the published paper, described significant differences between the biodynamic and the organic methodology; for instance, unspecified chemicals were added to the organic farm's compost, possibly leading to its slightly poorer performance.

- Biodynamic preparations increase the soil organic matter content.
- A further study investigated whether biodynamic preparations had any effect on the yield and growth of lentil and wheat crops, weed populations and soil fertility in the short term. The study found that "in general, soils and crops treated with biodynamic preparations showed few differences from those not treated". Plots tended with biodynamically treated compost produced results for yield, crop quality and soil fertility that were the same to those tended with non-biodynamic composts and NPK fertilizers. Some alteration was observed in the nitrogenous chemistry of the soil and grain where biodynamic field sprays were applied, however the study did not ascribe or discern any biological significance to the difference. Among the variables considered by the study, some measured outcomes correlated with biodynamic field spray usage, including a higher per-unit biomass yield ratio for lentils and a lowering of carbon and crude protein contents in wheat grains. The study's conclusion remarked that "any additional short-term benefits from biodynamic preparations remain questionable."
- Minor effects for the field sprays on the carbon content and soil microbial fatty acid profile, but no effects for the compost preparations
- A 1993 study compared soil quality and financial performance of Biodynamic and conventional farms in New Zealand. The study reported that, "The Biodynamic farms proved in most enterprises to have soils of higher biological and physical quality:

significantly greater in organic matter, content and microbial activity, more earthworms, better soil structure, lower bulk density, easier penetrability, and thicker topsoil." The biodynamic farms were just as financially viable on a per hectare basis. The study compared biodynamic farms with adjacent conventional farms, but didn't attempt to compare farms of similar size, or with similar crops.

Criticism

Biodynamic agriculture has been criticized as pseudoscience by scholars. In a 2002 newspaper editorial, Peter Treue, a researcher with the University of Kiel, characterised biodynamics as pseudoscience and argued that similar or equal results can be obtained using standard organic farming principles. He wrote that the biodynamic preparations more resemble alchemy or magic akin to geomancy.

In a 1994 analysis, Holger Kirchmann, a soil researcher with the Swedish University of Agricultural Sciences, concluded that Steiner's instructions were occult and dogmatic, and cannot contribute to the development of alternative or sustainable agriculture and that many of Steiner's statements are not provable because scientifically clear hypotheses cannot be made from his descriptions (for example, it is hard to prove that one has harnessed "cosmic forces" in the foods). Kirchmann asserted that when methods of biodynamic agriculture were tested scientifically, the results were unconvincing.

Further, in a 2004 overview of biodynamic agriculture, Linda Chalker-Scott, a researcher at Washington State University, characterised biodynamics as pseudoscience, writing that Steiner did not use scientific methods to formulate his theory of biodynamics, and that the later addition of valid organic farming techniques has "muddled the discussion" of Steiner's original idea. Based on the scant scientific testing of biodynamics, Chalker-Scott concluded "no evidence exists" that homeopathic preparations improve the soil.

In Michael Shermer's two-volume work, *The Skeptic Encyclopedia of Pseudoscience*, biodynamic agriculture is described as relying on astrological conditions, cosmic influences and magical rituals.

Biodynamic Wine

Biodynamic wines are wines made using the principles of biodynamic agriculture. Biodynamic refers to both the agricultural methods and the handling and processing of the fruit post-harvest.

Biodynamic Viticulture

The practice of biodynamics in viticulture (grape growing) has become popular in recent years in several growing regions, including France, Switzerland, Italy, Austria, Germany, Australia, Chile, South Africa, Canada, and the United States. A number of very high-end, high-profile commercial growers have converted recently to biodynamic practices. According to an article in *Fortune*, many of the top estates in France, "including Domaine Leroy in Burgundy, Château de la Roche-aux-Moines in the Loire, Maison Chapoutier in the Rhone Valley, and Domaine Zind-Humbrecht in Alsace," follow biodynamic viticulture. There are currently more than 450 biodynamic wine producers worldwide. Currently, for a wine to be labelled "biodynamic" it has to meet the stringent standards laid down by the Demeter Association, which is an internationally recognised certifying body.

Like biodynamic agriculture in general, biodynamic viticulture stems from the ideas and suggestions of Rudolf Steiner (1861–1925), who gave his now famous *Agriculture Course* in 1924, predating most of the organic movement. The principles and practices of biodynamics are based on his spiritual/practical philosophy, called anthroposophy, which includes understanding the ecological, the energetic, and the spiritual in nature.

As a practical method of farming, biodynamics embodies the ideal of ever-increasing ecological self-sufficiency just as with modern agro-ecology, but includes ethical-spiritual considerations. This type of viticulture views the farm as a cohesive, interconnected living system.

Efficacy

Some grape growers who have adopted biodynamic methods claim to have achieved improvements in the health of their vineyards, specifically in the areas of biodiversity, soil fertility, crop nutrition, and pest, weed, and disease management. For example, Anne-Claude Leflaive of Domaine Leflaive estate in Burgundy claims that the use of biodynamic methods saved a badly diseased vineyard, to the point that it now produces some of her most highly prized wines. A long-term study of one California luxury wine-farm found that improved quality for both biodynamic and organic could not be explained. This study in different vineyard blocks at a commercial vineyard in Ukiah, California found no difference between biodynamic methods with general organic farming methods with respect to soil quality, nor in the yield per vine, clusters per vine, and cluster and berry weight. However, one of the authors, Leo McCloskey has made the case that consumer quality scores, 100-point scores, are expected to be higher for both biodynamic and organic over traditional farming.

Biodynamic winemakers claim to have noted stronger, clearer, more vibrant tastes, as well as wines that remain drinkable longer. Biodynamic wines are more "floral", according to Spanish biodynamic vintner Perez Palacios. Biodynamic producers also note that their methods tend to result in better balance in growth, where the sugar production in the grapes coincides with physiological ripeness, resulting in a wine with the correct balance of flavour and alcohol content, even with changing climate conditions.

In a blind tasting of 10 pairs of biodynamic and conventionally-made wines, conducted by *Fortune* and judged by seven wine experts including a Master of Wine and head sommeliers, nine of the biodynamic wines were judged superior to their conventional counterpart. The biodynamic wines "were found to have better expressions of *terroir*, the way in which a wine can represent its specific place of origin in its aroma, flavour, and texture." Critics caution that such comparisons of wines of the same type need to be controlled for differences in soil and subsoil, and the farming and processing techniques used.

Critics acknowledge the high quality of biodynamic wines, but question whether many of the improvements in vineyard health and wine taste would have happened anyway if organic farming were used, without the mysticism and increased effort involved in biodynamics. Other critics attribute the success of biodynamic viticulture to the winemakers' higher craftsmanship and meticulous attention to detail. Ray Isle, managing editor of *Wine & Spirit* magazine, says, "So what if they also think burying cow horns full of manure will help them channel new life forces from the cosmos?"

Biodynamic Preparations

For a vineyard to be considered biodynamic the wine-grower must use the nine biodynamic preparations, as described in 1924 by Rudolf Steiner. These are made from cow manure, quartz (silica) and seven medicinal plants. Some of these materials are first transformed using animal organs as sheaths (NB: the animal organs are not used on the vineyards). Of the nine biodynamic preparations three are used as sprays (horn manure, horn silica and common horsetail) and the other six are applied to the vineyard via solid compost:

- Preparation 500 - Cow manure is buried in cow horns in the soil over winter. The horn is then dug up, its contents (called horn manure or '500') are then stirred in water and sprayed on the soil in the afternoon. The horn may be reused as a sheath.

- Preparation 501 - Ground quartz is buried in cow horns in the soil over summer. The horn is then dug up, its contents (called horn silica or '501') are then stirred in water and sprayed over the vines at daybreak. The horn may be reused as a sheath.
- Preparation 502 - Yarrow flowers are buried sheathed in a stag's bladder. This is hung in the summer sun, buried over winter, then dug up the following spring. The bladder's contents are removed and inserted in the compost (the used bladder is discarded).
- Preparation 503 - Chamomille, the German chamomile (*Matricaria chamomila*) flowers are sheathed in a cow intestine. This is hung in the summer sun, buried over winter, then dug up the following spring. The intestine's contents are removed and inserted in the compost (the used intestine is discarded).
- Preparation 504 - Stinging nettles are buried in the soil (with no animal sheath) in summer, are dug up the following autumn and are inserted in the compost.
- Preparation 505 - Oak bark is buried sheathed in the skull of a farm animal, the skull is buried in a watery environment over winter, then dug up. The skull's contents are removed and inserted in the compost (the used skull is discarded).
- Preparation 506 - Dandelion flowers are buried sheathed in a cow mesentery (peritoneum). This is hung in the summer sun, buried over winter, then dug up the following spring. The mesentery's contents are removed and inserted in the compost (the used mesentery is discarded).
- Preparation 507 - Valerian flower juice is sprayed over and/or inserted into the compost.
- Preparation 508 - Common Horsetail (*Equisetum arvense*) made either as a fresh tea or as a fermented liquid manure is applied either to the vines (in this case usually as a tea) or to the soil (in this case usually as a liquid manure).

Organic Wine

The most widely accepted definition of Organic wine is wine made from grapes grown in accordance with principles of organic farming, which typically excludes the use of artificial chemical fertilizers, pesticides, fungicides and herbicides.

Organic wine is generally consumed for its perceived health benefits and reduced environmental impact. The consumption of organic wine grew at a rate of 3.7 percent over the year ending September 19, 2009, out-pacing growth in the consumption of non-organic wine which grew 2% during a similar period. There are an estimated 1500-2000 organic wine producers globally, including negociant labels, with more than 885 of these organic domaines in France alone.

The legal definition of Organic Wine is a complex issue and varies from country to country. The primary difference in the way that organic wine is defined relates to the use (or non use) of preservatives during the wine-making process.

Production & Preservatives (as Related to Organic Wine)

Wine production comprises two main phases - that which takes place in the vineyard (i.e. grape growing) and that which takes place in the winery (i.e. fermentation of the grapes into wine, bottling etc.). The baseline definition of organic wine as "wine made with grapes farmed organically", deals only with the first phase (grape growing). There are numerous potential inputs which can be made during the second phase of production in order to ferment and preserve the wine. The most universal wine preservative is sulphur dioxide. The issue of wine preservation is central to the discussion of how organic wine is defined.

Wine matures over time, and it is widely considered that certain types of wines improve with aging, as the flavours become more integrated and balanced. As a result the greatest percentage of wines are produced in a way that allows them to last, sometimes as long as decades. The use of added sulfites is debated heavily within the organic winemaking community. Many vintners favour their use in small quantities for stabilisation of wine, while others frown on them. Currently the only effective preservatives that allow wines to last for a long period are 'non-organic'. While there are a growing number of producers making wine without added preservatives, it is generally acknowledged that these wines are for consumption within a few years of bottling.

The various legal definitions of organic wine serve to address this challenge regarding the use of preservatives. In some parts of the world, namely Europe, wine cannot legally be labelled as 'Organic Wine'. Wines that have added sulfites, but are otherwise organic, are labelled "wine made from organic grapes.". In the United States, wines certified "organic" under the National Organic Program cannot contain added sulfites.

Organic vs. Conventional Wine Growing

In conventional agriculture, chemical fertilizers are used to promote larger yields and protect against disease. These same chemicals are absorbed through the roots into the vine's sap and are then passed through leaves, and stems into the fruit. As a result, residues of these chemicals find their way into the finished wine. In addition to the effects of direct consumption, conventional 'chemical based' farming has a significant impact on soil and water quality. Wine growers farming with chemicals have to wear 'hazmat' style clothing and breathing apparatus to protect themselves while crop spraying.

Proponents of organically grown wine believe that chemical farming destroys the uniqueness of the land, and the unique flavour that this 'terroir' imparts to the wine.

In the USA, strict rules govern the organic winemaking process at all stages of production including harvesting, the types of yeasts that can be used during fermentation as well as storage conditions. These rules are applied for all imported and domestic wines that acquire USDA certification. In the USA, the total sulfite level must be less than 20 parts per million in order to receive organic certification.

Organic Certification

Organic certification of wine is complex; different nations have different certification criteria. In the United States, the National Organic Program, run by the United States Department of Agriculture, sets standards for certification of organic foods, including organic wines.

Some wineries that are technically organic choose not to be certified for various reasons.

Natural Wine

Organic wine refers to the absence of chemicals in the growth of grapes and in the winemaking facility and production process. It is a legal certification with third party inspection.

Natural winemaking is a style of winemaking that can be applied to any wine. It is loosely defined as using native yeasts in the fermentation process and minimal or no sulfur dioxide in the winemaking process. It may also mean unfined and unfiltered as well.

Natural winemaking is not governed by laws (at least not in the U.S.) and has no inspection or verification process (unless it is a biodynamic wine). Estimates are that less than 10 percent of the organically grown wine in the United States is made in a natural winemaking style, most of which is certified biodynamic wine.

Natural winemakers may or may not use organic or biodynamic grapes in their wines. Using native yeasts and relying on minimal manipulation often means that wines have a more unique profile from year to year. Different vintages vary more than conventionally made wine because of the non-interventionist approach. This is a key part of the natural wine aesthetic which emphasizes the least amount of intervention to bring the true flavour of each vintage to the glass.

The natural wine movement has grown in popularity in response to what some observers have called "Parkerization" or the globalization of wine tasting. A small number of critics' palates and the points system has come to define the market value of wines. The effect of this on wine producers has been to try to manipulate the taste of their wine (for example trying to increase the intensity of fruit and oak) in order to please certain wine critics and get higher ratings. As a result critics of these critics say this is causing an increasing uniformity amongst wines and a loss of regional and varietal character. The natural wine movement is one response to the global commodification of winemaking.

Sustainable Wines

Some farmers take additional steps beyond standard organic winemaking to apply sustainable farming practices. Examples include the use of composting and the cultivation of plants that attract insects that are beneficial to the health of the vines. Sustainable practices in these vineyards also extend to actions that have seemingly little or nothing to do with the production of grapes such as providing areas for wildlife to prevent animals from eating the grapes and allowing weeds and wildflowers to grow between the vines. Sustainable farmers may use bio-diesel for tractors in the vineyards to reduce emissions among the vines, or plough with horses.

Certified Naturally Grown

Certified Naturally Grown is a non-profit alternate farm assurance certification program created for small-scale organic farmers, and striving to strengthen the organic movement by preserving high organic standards and removing financial barriers that tend to exclude smaller farms that are selling locally and directly to their customers. The program is operated in the United States by a non-profit corporation, Certified Naturally Grown, Inc, based in Stone Ridge, New York. It was founded by Kate & Ron Khosla who operates a community supported farm in New Paltz, New York using the Community Supported Agriculture model. A United Kingdom sister organisation, the Wholesome Food Association, promotes similar food production standards, but does not operate a farm assurance or certification program.

To be Certified Organic (as opposed to Certified Naturally Grown) in the US, a grower must keep detailed records of planting, cultivation, fertilization, harvest, and storage, and must pay for both organisation membership and periodic inspection. This process works well for large-scale commercial growers, but becomes onerous for small mixed-agriculture farms. Since only certified seeds may be used, the varieties available to be grown are limited, and sustainable practices such as seed-saving is not permitted, unless the farmer also applies to be certified as a seed supplier.

Certified Naturally Grown farmers follow the USDA standards of the National Organic Program, but the record keeping and inspection process is tailored to accommodate the needs of small-scale mixed-agriculture farmers, and are not normally permitted to use the word "organic". Farmers commit to act as inspectors. Farmer-Inspectors are uniquely qualified to observe and note whether their neighbours are sticking to the standards, and are encouraged to provide helpful feedback, which helps foster a sense of community and sharing. Inspection forms are posted on the Internet for anytime public access, and all farms are subject to random pesticide residue testing. All in all, the CNG procedure requires significantly less paperwork, yet arguably results in more transparency and fostering of better farming practices, than the Certified Organic process, which primarily depends on farmer declarations backed by copious paperwork, and which inspects the paperwork rather than the farm.

A nationally recognised and endorsed grassroots effort, CNG exists because of enormous volunteer efforts, running primarily on free-will donations from farmers and supporters.

Industrial Agriculture

Industrial farming is a form of modern farming that refers to the industrialised production of livestock, poultry, fish, and crops. The methods of industrial agriculture are technoscientific, economic, and political. They include innovation in agricultural machinery and farming methods, genetic technology, techniques for achieving economies of scale in production, the creation of new markets for consumption, the application of patent protection to genetic information, and global trade. These methods are widespread in developed nations and increasingly prevalent worldwide. Most of the meat, dairy, eggs, fruits, and vegetables available in supermarkets are produced using these methods of industrial agriculture.

Historical Development and Future Prospects

The birth of industrial agriculture more or less coincides with that of the Industrial Revolution in general. The identification of nitrogen, potassium, and phosphorus (referred to by the acronym NPK) as critical factors in plant growth led to the manufacture of synthetic fertilizers, making possible more intensive types of agriculture. The discovery of vitamins and their role in animal nutrition, in the first two decades of the 20th century, led to vitamin supplements, which in the 1920s allowed certain livestock to be raised indoors, reducing their exposure to adverse natural elements. The discovery of antibiotics and vaccines facilitated raising livestock in concentrated, controlled animal feed operations by reducing diseases caused by crowding. Chemicals developed for use in World War II gave rise to synthetic pesticides. Developments in shipping networks and technology have made long-distance distribution of agricultural produce feasible.

Agricultural production across the world doubled four times between 1820 and 1975 to feed a global population of one billion human beings in 1800 and 6.5 billion in 2002. During the same period, the number of people involved in farming dropped as the process became more automated. In the 1930s, 24 percent of the American population worked in agriculture compared to 1.5 percent in 2002; in 1940, each farm worker supplied 11 consumers, whereas in 2002, each worker supplied 90 consumers. The number of farms has also decreased, and their ownership is more concentrated. In the U.S., four companies kill 81 percent of cows, 73 percent of sheep, 57 percent of pigs, and produce 50 percent of chickens, cited as an example of "vertical integration" by the president of the U.S. National Farmers' Union. In 1967, there were one million pig farms in America; as of 2002, there were 114,000, with 80 million pigs (out of 95 million) killed each year on factory farms, according to the U.S. National Pork Producers

Council. According to the Worldwatch Institute, 74 percent of the world's poultry, 43 percent of beef, and 68 percent of eggs are produced this way.

According to Denis Avery of the agribusiness funded Hudson Institute, Asia increased its consumption of pork by 18 million tons in the 1990s. As of 1997, the world had a stock of 900 million pigs, which Avery predicts will rise to 2.5 billion pigs by 2050. He told the College of Natural Resources at the University of California, Berkeley that three billion pigs will thereafter be needed annually to meet demand. He writes: "For the sake of the environment, we had better hope those hogs are raised in big, efficient confinement systems."

British Agricultural Revolution

The British agricultural revolution describes a period of agricultural development in Britain between the 16th century and the mid-19th century, which saw a massive increase in agricultural productivity and net output. This in turn supported unprecedented population growth, freeing up a significant percentage of the workforce, and thereby helped drive the Industrial Revolution. How this came about is not entirely clear. In recent decades, historians cited four key changes in agricultural practices, enclosure, mechanization, four-field crop rotation, and selective breeding, and gave credit to a relatively few individuals.

Challenges and Issues

The challenges and issues of industrial agriculture for global and local society, for the industrial agriculture sector, for the individual industrial agriculture farm, and for animal rights include the costs and benefits of both current practices and proposed changes to those practices. This is a continuation of thousands of years of the invention and use of technologies in feeding ever growing populations.

[W]hen hunter-gatherers with growing populations depleted the stocks of game and wild foods across the Near East, they were forced to introduce agriculture. But agriculture brought much longer hours of work and a less rich diet than hunter-gatherers enjoyed. Further population growth among shifting slash-and-burn farmers led to shorter fallow periods, falling yields and soil erosion. Plowing and fertilizers were introduced to deal with these problems - but once again involved longer hours of work and degradation of soil resources (Boserup, The Conditions of Agricultural Growth, Allen and Unwin, 1965, expanded and updated in Population and Technology, Blackwell, 1980.).

While the point of industrial agriculture is lower cost products to create greater productivity thus a higher standard of living as measured by available goods and services, industrial methods have side effects both good and bad. Further, industrial agriculture is not some single indivisible thing, but instead is composed of numerous separate elements, each of which can be modified, and in fact is modified in response to market conditions, government regulation, and scientific advances. So the question then becomes for each specific element that goes into an industrial agriculture method or technique or process: What bad side effects are bad enough that the financial gain and good side effects are outweighed? Different interest groups not only reach different conclusions on this, but also recommend differing solutions, which then become factors in changing both market conditions and government regulations.

Society

The major challenges and issues faced by society concerning industrial agriculture include:

Maximising the benefits:

- Cheap and plentiful food
- Convenience for the consumer
- The contribution to our economy on many levels, from growers to harvesters to processors to sellers.

while minimising the downsides:

- Environmental and social costs
- Damage to fisheries
- Cleanup of surface and groundwater polluted with animal waste
- Increased health risks from pesticides
- Increased ozone pollution via methane byproducts of animals
- Global warming from heavy use of fossil fuels.

Population Frowth

Very roughly:

- 30,000 years ago hunter-gatherer behaviour fed 6 million people
- 3,000 years ago primitive agriculture fed 60 million people
- 300 years ago intensive agriculture fed 600 million people
- Today industrial agriculture attempts to feed 6 billion people

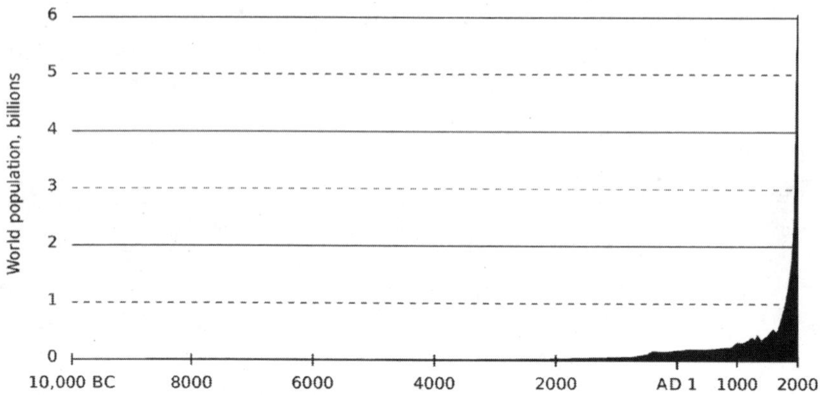

Figure 2: Population (est.) 10,000 BCE – 2000 CE.

Table 1: Estimated world o ulation at various dates, in thousands

Estimated world population at various dates, in **thousands**								
Year	**World**	**Africa**	**Asia**	**Europe**	**Central & South America**	**North America***	**Oceania**	**Notes**
8000 BCE	8 000							[11]
1000 BCE	50 000							[11]
500 BCE	100 000							[11]
1 CE	200,000 plus							[12]
1000	310 000							
1750	791 000	106 000	502 000	163 000	16 000	2 000	2 000	
1800	978 000	107 000	635 000	203 000	24 000	7 000	2 000	
1850	1 262 000	111 000	809 000	276 000	38 000	26 000	2 000	
1900	1 650 000	133 000	947 000	408 000	74 000	82 000	6 000	
1950	2 518 629	221 214	1 398 488	547 403	167 097	171 616	12 812	
1955	2 755 823	246 746	1 541 947	575 184	190 797	186 884	14 265	
1960	2 981 659	277 398	1 674 336	601 401	209 303	204 152	15 888	
1965	3 334 874	313 744	1 899 424	634 026	250 452	219 570	17 657	
1970	3 692 492	357 283	2 143 118	655 855	284 856	231 937	19 443	
1975	4 068 109	408 160	2 397 512	675 542	321 906	243 425	21 564	
1980	4 434 682	469 618	2 632 335	692 431	361 401	256 068	22 828	
1985	4 830 979	541 814	2 887 552	706 009	401 469	269 456	24 678	
1990	5 263 593	622 443	3 167 807	721 582	441 525	283 549	26 687	
1995	5 674 380	707 462	3 430 052	727 405	481 099	299 438	28 924	
2000	6 070 581	795 671	3 679 737	727 986	520 229	315 915	31 043	
2005	6 453 628	887 964	3 917 508	724 722	558 281	332 156	32 998**	

An example of industrial agriculture providing cheap and plentiful food is the U.S.'s "most successful program of agricultural development of any country in the world". Between 1930 and 2000 U.S. agricultural productivity (output divided by all inputs) rose by an average of about 2 percent annually causing food prices paid by consumers to decrease. "The percentage of U.S. disposable income spent on food prepared at home decreased, from 22 percent as late as 1950 to 7 percent by the end of the century."

Liabilities

Environment: Industrial agriculture uses huge amounts of water, energy, and industrial chemicals; increasing pollution in the arable land, usable water and atmosphere. Herbicides, insecticides, fertilizers, and animal waste products are accumulating in ground and surface waters. "Many of the negative effects of industrial agriculture are remote from fields and farms. Nitrogen compounds from the Midwest, for example, travel down the Mississippi to degrade coastal fisheries in the Gulf of Mexico. But other adverse effects are showing up within agricultural production systems — for example, the rapidly developing resistance among pests is rendering our arsenal of herbicides and insecticides increasingly ineffective."

Social

A study done for the US. Office of Technology Assessment conducted by the UC Davis Macrosocial Accounting Project concluded that industrial agriculture is associated with substantial deterioration of human living conditions in nearby rural communities.

Animals

"Confined animal feeding operations" or "intensive livestock operations", can hold large numbers (some up to hundreds of thousands) of animals, often indoors. These animals are typically cows, hogs, turkeys, or chickens. The distinctive characteristics of such farms is the concentration of livestock in a given space. The aim of the operation is to produce as much meat, eggs, or milk at the lowest possible cost and with the greatest level of food safety.

Food and water is supplied in place, and artificial methods are often employed to maintain animal health and improve production, such as therapeutic use of antimicrobial agents, vitamin supplements and growth hormones. Growth hormones are not used in chicken meat production nor are they used in the European Union for any animal. In meat production, methods are also sometimes employed to control undesirable behaviours often related to stresses of being confined in restricted areas with other animals. More docile breeds are sought (with natural dominant behaviours bred out for example), physical restraints to stop interaction, such as individual cages for chickens, or animals physically modified, such as the debeaking of chickens to reduce the harm of fighting. Weight gain is encouraged by the provision of plentiful supplies of food to animals breed for weight gain.

The designation "confined animal feeding operation" in the U.S. resulted from that country's 1972 Federal Clean Water Act, which was enacted to protect and restore lakes and rivers to a "fishable, swimmable" quality. The United States Environmental Protection Agency (EPA) identified certain animal feeding operations, along with many other types of industry, as point source polluters of groundwater. These operations were designated as CAFOs and subject to special anti-pollution regulation.

In 24 states in the U.S., isolated cases of groundwater contamination has been linked to CAFOs. For example, the ten million hogs in North Carolina generate 19 million tons of waste per year. The U.S. federal government acknowledges the waste disposal issue and requires that animal waste be stored in lagoons. These lagoons can be as large as 7.5 acres (30,000 m^2). Lagoons not protected with an impermeable liner can leak waste into groundwater under some conditions, as can runoff from manure spread back onto fields as fertilizer in the case of an unforeseen heavy rainfall. A lagoon that burst in 1995 released 25 million gallons of nitrous sludge in North Carolina's New River. The spill allegedly killed eight to ten million fish.

The large concentration of animals, animal waste, and dead animals in a small space poses ethical issues to some consumers. Animal rights and animal welfare activists have charged that intensive animal rearing is cruel to animals. As they become more common, so do concerns about air pollution and ground water contamination, and the effects on human health of the pollution and the use of antibiotics and growth hormones.

According to the U.S. Centres for Disease Control and Prevention (CDC), farms on which animals are intensively reared can cause adverse health reactions in farm workers. Workers may develop acute and chronic lung disease, musculoskeletal injuries, and may catch infections that transmit from animals to human beings. These type of transmissions, however, and extremely rare, as zoonotic diseases are uncommon.

Crops

The projects within the Green Revolution spread technologies that had already existed, but had not been widely used outside of industrialised nations. These technologies included pesticides, irrigation projects, and synthetic nitrogen fertilizer.

The novel technological development of the Green Revolution was the production of what some referred to as "miracle seeds." Scientists created strains of maize, wheat, and rice that are generally referred to as HYVs or "high-yielding varieties." HYVs have an increased nitrogen-absorbing potential compared to other varieties. Since cereals that absorbed extra nitrogen would typically lodge, or fall over before harvest, semi-dwarfing genes were bred into their genomes. Norin 10 wheat, a variety developed by Orville Vogel from Japanese dwarf wheat varieties, was instrumental in developing Green Revolution wheat cultivars. IR8, the first widely implemented HYV rice to be developed by the International Rice Research Institute, was created through a cross between an Indonesian variety named "Peta" and a Chinese variety named "Dee Geo Woo Gen."

With the availability of molecular genetics in Arabidopsis and rice the mutant genes responsible (*reduced height(rht), gibberellin insensitive (gai1)* and *slender rice (slr1)*) have been cloned and identified as cellular signalling components of gibberellic acid, a phytohormone involved in regulating stem growth via its effect on cell division. Stem growth in the mutant background is significantly reduced leading to the dwarf phenotype. Photosynthetic investment in the stem is reduced dramatically as the shorter plants are inherently more stable mechanically. Assimilates become redirected to grain production, amplifying in particular the effect of chemical fertilizers on commercial yield.

HYVs significantly outperform traditional varieties in the presence of adequate irrigation, pesticides, and fertilizers. In the absence of these inputs, traditional varieties may outperform HYVs. One criticism of HYVs is that they were developed as F1 hybrids, meaning they need to be purchased by a farmer every season rather than saved from previous seasons, thus increasing a farmer's cost of production.

Sustainable Agriculture

The idea and practice of sustainable agriculture has arisen in response to the problems of industrial agriculture. Sustainable agriculture integrates three main goals: environmental stewardship, farm profitability, and prosperous farming communities. These goals have been defined by a variety of disciplines and may be looked at from the vantage point of the farmer or the consumer.

Organic Farming Methods

Organic farming methods combine some aspects of scientific knowledge and highly limited modern technology with traditional farming practices; accepting some of the methods of industrial agriculture while rejecting others. Organic methods rely on naturally occurring biological processes, which often take place over extended periods of time, and a holistic approach; while chemical-based farming focuses on immediate, isolated effects and reductionist strategies.

Integrated Multi-Trophic Aquaculture is an example of this holistic approach. Integrated Multi-Trophic Aquaculture (IMTA) is a practice in which the by-products (wastes) from one species are recycled to become inputs (fertilizers, food) for another. Fed aquaculture (e.g. fish, shrimp) is combined with inorganic extractive (e.g. seaweed) and organic extractive (e.g. shellfish) aquaculture to create balanced systems for environmental sustainability (biomitigation), economic stability (product diversification and risk reduction) and social acceptability (better management practices).

Integrated Multi-trophic Aquaculture

Integrated multi-trophic aquaculture (IMTA) provides the by-products, including waste, from one aquatic species as inputs (fertilizers, food) for another. Farmers combine fed aquaculture (e.g., fish, shrimp) with inorganic extractive (e.g., seaweed) and organic extractive (e.g., shellfish) aquaculture to create balanced systems for environment remediation (biomitigation), economic stability (improved output, lower cost, product diversification and risk reduction) and social acceptability (better management practices).

Selecting appropriate species and sizing the various populations to provide necessary ecosystem functions allows the biological and chemical processes involved to achieve a stable balance, mutually benefiting the organisms and improving ecosystem health.

Ideally, the co-cultured species each yield valuable commercial "crops". IMTA can synergistically increase total output, even if some of the crops yield less than they would, short-term, in a monoculture.

Terminology and Related Approaches

"Integrated" refers to intensive and synergistic cultivation, using water-born nutrient and energy transfer. "Multi-trophic" means that the various species occupy different trophic levels, i.e., different (but adjacent) links in the food chain.

IMTA is a specialised form of the age-old practice of aquatic polyculture, which was the co-culture of various species, often without regard to trophic level. In this broader case, the organisms may share biological and chemical processes that are minimally complementary, potentially leading to significant ecosystem shifts/damage. Some traditional systems did culture species that occupied multiple niches within the same pond, but with limited intensity and management.

The more general term "Integrated Aquaculture" is used to describe the integration of monocultures through water transfer. The terms "IMTA" and "integrated aquaculture" differ primarily in their precision and are sometimes interchanged. Aquaponics, fractionated aquaculture, IAAS (integrated agriculture-aquaculture systems), IPUAS (integrated peri-urban-aquaculture systems), and IFAS (integrated fisheries-aquaculture systems) are variations on the IMTA concept.

Range of Approaches

Today, low intensity traditional/incidental multi-trophic aquaculture is much more common than modern IMTA. Most are relatively simple, such as fish/seaweed/shellfish.

True IMTA can be land-based, using ponds or tanks, or open-water marine or freshwater systems. Implementations have included species combinations such as shellfish/shrimp, fish/seaweed/shellfish, fish/shrimp and seaweed/shrimp.

In the future, systems with other components for additional functions, or similar functions but different size brackets of particles, are likely. Multiple regulatory issues remain open.

Modern History of Land-based Systems

Ryther and co-workers created modern, integrated, intensive, land mariculture. They originated, both theoretically and experimentally, the integrated use of extractive organisms—shellfish, microalgae and seaweeds—in the treatment of household effluents, descriptively and with quantitative results. A domestic wastewater effluent, mixed with seawater, was the nutrient source for phytoplankton, which in turn became food for oysters and clams. They cultivated other organisms in a food chain rooted in the farm's organic sludge. Dissolved nutrients in the final effluent were filtered by seaweed (mainly Gracilaria and Ulva) biofilters. The value of the original organisms grown on human waste effluents was minimal.

In 1976, Huguenin proposed adaptations to the treatment of intensive aquaculture effluents in both inland and coastal areas. Tenore followed by integrating with their system of carnivorous fish and the macroalgivore abalone.

In 1977, Hughes-Games described the first practical marine fish/shellfish/phytoplankton culture, followed by Gordin, et al., in 1981. By 1989, a semi-intensive (1 kg fish/m^{-3}) seabream and grey mullet pond system by the Gulf of Aqaba (Eilat) on the Red Sea supported dense diatom populations, excellent for feeding oysters. Hundreds of kilos of fish and oysters cultured here were sold. Researchers also quantified the water quality parameters and nutrient budgets in (5 kg fish m^{-3}) green water seabream ponds. The phytoplankton generally maintained reasonable water quality and converted on average over half the waste nitrogen into algal biomass. Experiments with intensive bivalve cultures yielded high bivalve growth rates. This technology supported a small farm in southern Israel.

Sustainability

IMTA promotes economic and environmental sustainability by converting byproducts and uneaten feed from fed organisms into harvestable crops, thereby reducing eutrophication, and increasing economic diversification.

Properly managed multi-trophic aquaculture accelerates growth without detrimental side-effects. This increases the site's ability to assimilate the cultivated organisms, thereby reducing negative environmental impacts.

IMTA enables farmers to diversify their output by replacing purchased inputs with byproducts from lower trophic levels, often without new sites. Initial economic research suggests that IMTA can increase profits and can reduce financial risks due to weather, disease and market fluctuations. Over a dozen studies have investigated the economics of IMTA systems since 1985.

Nutrient Flow

Typically, carnivorous fish or shrimp occupy IMTA's higher trophic levels. They excrete soluble ammonia and phosphorus (orthophosphate). Seaweeds and similar species can extract these inorganic nutrients directly from their environment. Fish and shrimp also release organic nutrients which feed shellfish and deposit feeders.

Species such as shellfish that occupy intermediate trophic levels often play a dual role, both filtering organic bottom-level organisms from the water and generating some ammonia. Waste feed may also provide additional nutrients; either by direct consumption or via decomposition into individual nutrients.

Recovery Efficiency

Nutrient recovery efficiency is a function of technology, harvest schedule, management, spatial configuration, production, species selection, trophic level biomass ratios, natural food availability, particle size, digestibility, season, light, temperature, and water flow. Since these factors significantly vary by site and region, recovery efficiency also varies. In a hypothetical family-scale fish/microalga /bivalve/ seaweed farm, based on pilot scale data, at least 60% of nutrient input reached commercial products, nearly three times more than in modern net pen farms. Expected average annual yields of the system for a hypothetical 1 hectare (2.5 acres) were 35 tonnes (34 long tons; 39 short tons) of seabream, 100 tonnes (98 long tons; 110 short tons) of bivalves and 125 tonnes (123 long tons; 138 short tons) of seaweeds. These results required precise water quality control and attention to suitability for bivalve nutrition, due to the difficulty in maintaining consistent phytoplanton populations.

Seaweeds' nitrogen uptake efficiency ranges from 2-100% in land-based systems. Uptake efficiency in open-water IMTA is unknown.

Food Safety and Quality

Feeding the wastes of one species to another has the potential for contamination, although this has yet to be observed in IMTA systems. Mussels and kelp growing adjacent to Atlantic salmon cages in the Bay of Fundy have been monitored since 2001 for contamination by medicines, heavy metals, arsenic, PCBs and pesticides. Concentrations are consistently either non-detectable or well below regulatory limits established by the Canadian Food Inspection Agency, the United States Food and Drug Administration and European Community Directives. Taste testers indicate that these mussels are free of "fishy" taste and aroma and could not distinguish them from "wild" mussels. The mussels' meat yield is significantly higher, reflecting the increase in nutrient availability.

Selected Projects

Historic and ongoing research projects include:

Asia

Japan, China, South Korea, Thailand, Vietnam, Indonesia, etc. have co-cultured aquatic species for centuries in marine, brackish and fresh water environments. Fish, shellfish and seaweeds have been cultured together in bays, lagoons and ponds. Trial and error has improved integration over time. The proportion of Asian aquaculture production that occurs in IMTA systems is unknown.

After the 2004 tsunami, many of the shrimp farmers in Aceh Province of Indonesia and Ranong Province of Thailand were trained in IMTA. This has been especially important as the mono-culture of marine shrimp was widely recognised as unsustainable. Production of tilapia, mud crabs, seaweeds, milkfish, and mussels have been incorporated. AquaFish Collaborative Research Support Program

Canada

Bay of Fundy: Industry, academia and government are collaborating here to expand production to commercial scale. The current system integrates Atlantic salmon, blue mussels and kelp; deposit feeders are under consideration. AquaNet (one of Canada's Networks of Centres of Excellence) funded phase one. The Atlantic Canada Opportunities Agency is presently funding phase two. The project leaders are Thierry Chopin (University of New Brunswick in Saint John) and Shawn Robinson (Department of Fisheries and Oceans, St. Andrews Biological Station).

Pacific SEA-lab

Pacific SEA-lab is researching and is licensed for the co-culture of sablefish, scallops, oysters, blue mussels, urchins and kelp. "SEA" stands for Sustainable Ecological Aquaculture. The project presently aims to balance four species. The project is headed by Stephen Cross under a British Columbia Innovation Award at the University of Victoria Coastal Aquaculture Research & Training (CART) network.

Chile

The i-mar Research Center at the Universidad de Los Lagos, in Puerto Montt is working to reduce the environmental impact of intensive salmon culture. Initial research involved trout, oysters and seaweeds. Present research is focusing on open waters with salmon, seaweeds and abalone. The project leader is Alejandro Buschmann.

Israel

Sea Or Marine Enterprises Ltd.: SeaOr Marine Enterprises Ltd., which operated for several years on the Israeli Mediterranean coast, north of Tel Aviv, cultured marine fish (gilthead seabream), seaweeds (Ulva and Gracilaria) and Japanese abalone. Its approach leveraged local climate, and recycled fish waste products into seaweed biomass, which was fed to the abalone. It also effectively purified the water sufficiently to allow the water to be recycled to the fishponds and to meet point-source effluent environmental regulations.

PGP Ltd.

PGP Ltd. is a small farm in Southern Israel. It cultures marine fish, microalgae, bivalves and Artemia. Effluents from seabream and seabass collect in sedimentation ponds, where dense populations of microalgae—mostly diatoms—develop. Clams, oysters and sometimes Artemia filter the microalgae from the water, producing a clear effluent. The farm sells the fish, bivalves and Artemia.

South Africa

Three farms currently grow seaweeds for feed in abalone effluents in land-based tanks. Up to 50% of re-circulated water passes through the seaweed tanks. Somewhat uniquely, neither fish nor shrimp comprise the upper trophic species. The motivation is to avoid over-harvesting natural seaweed beds and red tides, rather than nutrient abatement. These commercial successes developed from research collaboration between Irvine and Johnson Cape Abalone and scientists from the University of Cape Town and the University of Stockholm.

United Kingdom

The Scottish Association for Marine Science, in Oban is developing co-cultures of salmon, oysters, sea urchins, and brown and red seaweeds via several projects. Research focuses on biological and physical processes, as well as production economics and implications for coastal zone management. Researchers include: M. Kelly, A. Rodger, L. Cook, S. Dworjanyn, and C. Sanderson.

Other Projects

Several other projects over the last three decades have aided IMTA development.

Aquaponics

Aquaponics is a sustainable food production system that combines a traditional aquaculture (raising aquatic animals such as fish, crayfish or prawns in tanks) with hydroponics (cultivating plants in water) in a symbiotic environment. In the aquaculture, effluents accumulate in the water, increasing toxicity for the fish. This water is led to a hydroponic system where the by-products from the aquaculture are filtered out by the plants as vital nutrients, after which the cleansed water is recirculated back to the animals. The term *aquaponics* is a portmanteau of the terms *aquaculture* and *hydroponic.*

Figure 3: A small, portable aquaponics system

Aquaponic systems vary in size from small indoor or outdoor units to large commercial units, using the same technology. The systems usually contain fresh water, but salt water systems are plausible depending on the type of aquatic animal and which plants. Aquaponic science may still be considered to be at an early stage.

Function

Aquaponics consists of two main parts, with the aquaculture part for raising aquatic animals and the hydroponics part for growing plants. Aquatic effluents resulting from uneaten feed or raising animals like fish, accumulates in water due to the closed system recirculation of most aquaculture systems. The effluent-rich water becomes toxic to the aquatic animal in high concentrations but these effluents are nutrients essential for plant growth. Although consisting primarily of these two parts, aquaponics system are usually grouped into several components or subsystems responsible for the effective removal of solid wastes, for adding bases to neutralise acids, or for maintaining water oxygenation. Typical components include:

- *Rearing tank*: the tanks for raising and feeding the fish;
- *Solids removal*: a unit for catching uneaten food and detached biofilms, and for settling out fine particulates;
- *Biofilter*: a place where the nitrification bacteria can grow and convert ammonia into nitrates, which are usable by the plants;

- *Hydroponics subsystem*: the portion of the system where plants are grown by absorbing excess nutrients from the water;
- *Sump*: the lowest point in the system where the water flows to and from which it is pumped back to the rearing tanks.

Depending on the sophistication and cost of the aquaponics system, the units for solids removal, biofiltration, and/or the hydroponics subsystem may be combined into one unit or subsystem, which prevents the water from flowing directly from the aquaculture part of the system to the hydroponics part.

Nitrification

Nitrification, the aerobic conversion of ammonia into nitrates, is one of the most important functions in an aquaponics system as it reduces the toxicity of the water for fish, and allows the resulting nitrate compounds to be removed by the plants for nourishment. Ammonia is steadily released into the water through the excreta and gills of fish as a product of their metabolism, but must be filtered out of the water since higher concentrations of ammonia (commonly between 0.5 and 1 ppm) can kill fish. Although plants can absorb ammonia from the water to some degree, nitrates are assimilated more easily, thereby efficiently reducing the toxicity of the water for fish. Ammonia can be converted into other nitrogenous compounds through healthy populations of:

- *Nitrosomonas*: bacteria that convert ammonia into nitrites, and
- *Nitrobacter*: bacteria that convert nitrites into nitrates.

In an aquaponics system, the bacteria responsible for this process form a biofilm on all solid surfaces throughout the system that are in constant contact with the water. The submerged roots of the vegetables combined have a large surface area, so that many bacteria can accumulate there. Together with the saliency of ammonia and nitrites in the water, the surface area determines the speed with which nitrification takes place. Care for these bacterial colonies is important as to regulate the full assimilation of ammonia and nitrite. This is why most aquaponics systems include a biofiltering unit, which helps facilitate growth of these microorganisms. Typically, after a system has stabilised ammonia levels range from 0.25 to 2.0 ppm; nitrite levels range from 0.25 to 1 ppm, and nitrate levels range from 2 to 150 ppm. During system startup, spikes may occur in the levels of ammonia (up to 6.0 ppm) and nitrite (up to 15 ppm), with nitrate levels peaking later in the startup phase. Since the nitrification process acidifies the water, non-sodium bases such as potassium hydroxide or calcium hydroxide can be added for neutralising the water's pH if insufficient quantities are naturally present in the water to provide a buffer against acidification. In addition, selected minerals or nutrients such as iron can be added in addition to the fish waste that serves as the main source of nutrients to plants.

A good way to deal with solids buildup in aquaponics is the use of worms, which liquefy the solid organic matter so that it can be utilised by the plants and/or animals.

Hydroponics Subsystem

Plants are grown as in hydroponics systems, with their roots immersed in the nutrient-rich effluent water. This enables them to filter out the ammonia that is toxic to the aquatic animals, or its metabolites. After the water has passed through the hydroponic subsystem, it is cleaned and oxygenated, and can return to the aquaculture vessels. This cycle is continuous. Common aquaponic applications of hydroponic systems include:

- *Deep-water raft aquaponics*: styrofoam rafts floating in a relatively deep aquaculture basin in troughs.
- *Recirculating aquaponics*: solid media such as gravel or clay beads, held in a container that is flooded with water from the aquaculture. This type of aquaponics is also known as *closed-loop aquaponics.*
- *Reciprocating aquaponics*: solid media in a container that is alternately flooded and drained utilising different types of siphon drains. This type of aquaponics is also known as *flood-and-drain aquaponics* or *ebb-and-flow aquaponics.*
- Other systems use towers that are trickle-fed from the top, nutrient film technique channels, horizontal PVC pipes with holes for the pots, plastic barrels cut in half with gravel or rafts in them. Each approach has its own benefits.

Most green leaf vegetables grow well in the hydroponic subsystem, although most profitable are varieties of chinese cabbage, lettuce, basil, roses, tomatoes, okra, cantaloupe and bell peppers. Other species of vegetables that grow well in an aquaponic system include beans, peas, kohlrabi, watercress, taro, radishes, strawberries, melons, onions, turnips, parsnips, sweet potato and herbs. Since plants at different growth stages require different amounts of minerals and nutrients, plant harvesting is staggered with seedings growing at the same time as mature plants. This ensures stable nutrient content in the water because of continuous symbiotic cleansing of toxins from the water.

Aquaculture Subsystem

Freshwater fish are the most common aquatic animal raised using aquaponics, although saltwater fish, crayfish and prawns may also be used. In practice, tilapia are the most popular fish for home and commercial projects that are intended to raise edible fish, although barramundi, Silver Perch, tandanus catfish, Jade perch and Murray cod are also used.

Normal Operations

Aquaponic systems do not typically discharge or exchange water under normal operation, but instead recirculate and reuse water very effectively. The system relies on the relationship between the animals and the plants to maintain a stable aquatic environment that experience a minimum of fluctuation in ambient nutrient and oxygen levels. Water is only added to replace water loss from absorption and transpiration by plants, evaporation into the air from surface water, overflow from the system from rainfall, and removal of biomass such as settled solid wastes from the system.

As a result, aquaponics uses approximately 2% of the water that a conventionally irrigated farm requires for the same vegetable production. This allows for aquaponic production of both crops and fish in areas where water or fertile land is scarce.

Aquaponic systems can also be used to replicate controlled wetland conditions that are useful for water treatment by reclaiming potable water from typical household sewage. The nutrient-filled overflow water can be accumulated in catchment tanks, and reused to accelerate growth of crops planted in soil, or it may be pumped back into the aquaponic system to top up the water level..

The three main inputs to the system are water, feed given to the aquatic animals, and electricity to pump water between the aquaculture subsystem and the hydroponics subsystem. Spawn or fry may be added to replace grown fish that are taken out from the system to retain a stable system.

In terms of outputs, an aquaponics system may continually yield plants such as vegetables grown in hydroponics, and edible aquatic species raised in an aquaculture.

History

Ancient: Aquaponics has ancient roots, although there is some debate on its first occurrence:

- Aztec cultivated agricultural islands known as *chinampas* and are considered by some as the first form of aquaponics for agricultural use, where plants were raised on stationary (and sometime movable) islands in lake shallows and waste materials dredged from the Chinampa canals and surrounding cities are used to manuallly irrigate the plants
- South China and Thailand who cultivated and farmed rice in paddy fields in combination with fish are cited as examples of early aquaponics. These polycultural farming systems existed in many Far Eastern countries and raised fish such as the oriental loach, swamp eel, Common and crucian carp as well as pond snails in the paddies.

Regions

North America:

United States: Inspired by the successes of the New Alchemy Institute and the North Carolina State University with aquaponics, other institutes followed suit. Besides the reciprocating aquaponics based on the techniques developed by Dr. Mark McMurtry et al. at the New Alchemy Institute and North Carolina State University, Dr. James Rakocy and his colleagues at the University of the Virgin Islands researched and developed the "Deep Water" or "Raft Culture" aquaponics The system combines tilapia with various vegetables.

In 1997 Rebecca L. Nelson and John S. Pade began publishing the Aquaponics Journal, a quarterly scientific journal that brings together research and various applications of aquaponics from around the globe. In 2008, they wrote and published the first comprehensive book on aquaponics, *Aquaponic Food Production.* Nelson and Pade work closely with Dr. Rakocy to bring the research on aquaponics into mainstream agriculture, and offer systems, supplies, training and consultancy professionally.

Recent years have seen a shift towards community integration of aquaponics, such as the nonprofit foundation Growing Power that offers Milwaukee youth job opportunities and training while growing food for their community. The model has spawned several satellite projects in other cities. In 2010, a partnership of experts in sustainable agriculture was formed under the name AquaPlanet to promote the technology through media and consulting. The team includes Dr. Mark McMurtry and Barrel-Ponics inventor Travis Hughey.

Canada: The first aquaponics research in Canada was a small system added onto existing aquaculture research at a research station in Lethbridge. Canada saw a rise in aquaponics setups throughout the 90s, predominantly as commercial installations, that for example combine trout with floating lettuce production, or to water fruiting vegetable crops that warm up the water too much to be recirculated back into the fish ponds. Eels are also known to be raised. A set-up based on the deep water system developed at the University of Virgin Islands was built in a greenhouse at Brooks, Alberta where Dr. Nick Savidov and colleagues researched aquaponics from a background of plant science. The team made findings on rapid root growth in aquaponics systems, on closing the solid waste loop, and that because of certain advantages in the system over traditional aquaculture, the system can run well at a low pH level, which is favoured by plants but not fish. The Edmonton Aquaponics Society in Northern Alberta is adapting Dr. Savidov's commercially sized system to a smaller scale prototype that can be operated by families, small groups, or restaurants. They intend to further develop the closed solid waste loop.

South America:

Barbados: Barbados is a densely populated island that deals with water scarcity. In 40 years' time, focus has shifted from domestic fruit and vegetable production on small farms to importing 80% of all fruits and vegetables for cost reasons. Aquaponics would make Barbados and other Caribbean islands less dependent on the world food market and reduce stress on the dwindling fish supplies. An inter-organisational project that started in late 2009 sets out to encourage and enable Barbadians to start aquaponics at home, with revenue generated by selling produce to tourists.

Asia:

Bangladesh: Bangladesh is the world's most densely populated country. Most of its farmers use/misuse agrochemicals to enhance food production/storage life. However, the country is yet to establish a good monitoring system to ensure chemical-residue free safe food in the market [23], capable city dwellers are looking for a better way to grow their own safe foods for domestic consumption. To help support mainly this client group and people living in adverse climatic conditions such as salinity-prone southern part and flood-prone haor area in the eastern region, a low-cost model of backyard aqaponics system[24] has been developed by Professor Dr. M.A. Salam at the Department of Aquaculture of Bangladesh Agricultural University, Mymensingh. The initial model has been developed in 2011, and demonstrated and reported to the press in January 2012. Further researches may be necessary to develop appropriate models for different situations among client groups.

Figure 4: Vegetable production part of the low-cost Backyard Aqauponics System developed at Bangladesh Agricultural University

Taiwan: Taiwan is a densely populated island that is faced with freshwater scarcity. Dispensation of water is government-controlled. The closed-loop system of aquaponics is used by agricultural farmers to save water by also rearing fish, while fish farmers grow plants that filter the water from the fish tanks.

Australia: Due to a ban on tilapia in all states except for Western Australia, native freshwater fish including silver perch, jade perch, sleepy cod, murray cod and barramundi are popular in aquaponics and aquaculture systems, along with non-native rainbow trout, brown trout and crayfish such as common yabby and redclaw.

Pros and Cons

The unique advantages of aquaponic systems are:

- Conservation through constant water reuse and recycling.
- Organic fertilization of plants with natural fish emulsion.
- The elimination of solid waste disposal from intensive aquaculture.
- The reduction of needed cropland to produce crops.
- The overall reduction of the environmental footprint of crop production.

- Building small efficient commercial installations near markets reduces food miles.
- Reduction of pathogens that often plague aquaculture production systems.

Some conceivable disadvantages with aquaponics are:

- Initial expenses for housing, tank, plumbing, pumps, and grow beds.
- The infinite number of ways in which a system can be configured lends itself to equally varying results, conflicting research, and successes or failures.
- Some aquaponic installations rely heavily on man-made energy, technological solutions, and environmental control to achieve recirculation and water/ambient temperatures. However, if a system is designed with energy conservation in mind, using alternative energy and a reduced number of pumps by letting the water flow downwards as much as possible, it can be highly energy efficient.
- While careful design can minimise the risk, aquaponics systems can have multiple 'single points of failure' where problems such as an electrical failure or a pipe blockage can lead to a complete loss of fish stock.
- Like all aquaculture based systems, stock feed usually consists of fish meal derived from lower value species. Ongoing depletion of wild fish stocks makes this practice unsustainable. Organic fish feeds may prove to be a viable alternative that negates this concern. Other alternatives include growing duckweed with an aquaponics system that feeds the same fish grown on the system, excess worms grown from vermiculture composting, as well as growing black soldier fly larvae to feed to the fish using composting grub growers.

Intercropping

Intercropping is the practice of growing two or more crops in proximity. The most common goal of intercropping is to produce a greater yield on a given piece of land by making use of resources that would otherwise not be utilised by a single crop. Careful planning is required, taking into account the soil, climate, crops, and varieties. It is particularly important not to have crops competing with each other for physical space, nutrients, water, or sunlight. Examples of intercropping strategies are planting a deep-rooted crop with a shallow-rooted crop, or planting a tall crop with a shorter crop that requires partial shade. Inga alley cropping has been proposed as an alternative to the ecological destruction of Slash-and-burn farming.

Figure 5: Pepper intercropped with coffee in Colombia's southwestern Cauca Department.

When crops are carefully selected, other agronomic benefits are also achieved. Lodging-prone plants, those that are prone to tip over in wind or heavy rain, may be given structural support by their companion crop (Trenbath 1976). Delicate or light sensitive plants may be given shade or protection, or otherwise wasted space can be utilised. An example is the tropical multi-tier system where coconut occupies the upper tier, banana the middle tier, and pineapple, ginger, or leguminous fodder, medicinal or aromatic plants occupy the lowest tier.

Intercropping of compatible plants also encourages biodiversity, by providing a habitat for a variety of insects and soil organisms that would not be present in a single-crop environment. This biodiversity can in turn help to limit outbreaks of crop pests (Altieri 1994) by increasing the diversity or abundance of natural enemies, such as spiders or parasitic wasps. Increasing the complexity of the crop environment through intercropping also limits the places where pests can find optimal foraging or reproductive conditions.

The degree of spatial and temporal overlap in the two crops can vary somewhat, but both requirements must be met for a cropping system to be an intercrop. Numerous types of intercropping, all of which vary the temporal and spatial mixture to some degree, have been identified (Andrews & Kassam 1976). These are some of the more significant types:

- Mixed intercropping, as the name implies, is the most basic form in which the component crops are totally mixed in the available space.
- Row cropping involves the component crops arranged in alternate rows. Variations include alley cropping, where crops are grown in between rows of trees, and strip cropping, where multiple rows, or a strip, of one crop are alternated with multiple rows of another crop.
- Intercropping also uses the practice of sowing a fast growing crop with a slow growing crop, so that the fast growing crop is harvested before the slow growing crop starts to mature. This obviously involves some temporal separation of the two crops.
- Further temporal separation is found in relay cropping, where the second crop is sown during the growth, often near the onset of reproductive development or fruiting, of the first crop, so that the first crop is harvested to make room for the full development of the second.

Chapter 4

Multiple Cropping

In agriculture, multiple cropping is the practice of growing two or more crops in the same space during a single growing season. It is a form of polyculture. It can take the form of double-cropping, in which a second crop is planted after the first has been harvested, or relay cropping, in which the second crop is started amidst the first crop before it has been harvested. A related practice, companion planting, is sometimes used in gardening and intensive cultivation of vegetables and fruits. One example of multi-cropping is tomatoes + onions + marigold; the marigolds repel some tomato pests.

Multiple cropping is found in many agricultural traditions. In the Garhwal Himalaya of India, a practice called *baranaja* involves sowing 12 or more crops on the same plot, including various types of beans, grams, and millets, and harvesting them at different times.

In the cultivation of rice, multiple cropping requires effective irrigation, especially in areas with a dry season. Rain that falls during the wet season permits the cultivation of rice during that period, but during the other half of the year, water cannot be channelled into the rice fields without an irrigation system. The Green Revolution in Asia led to the development of high-yield varieties of rice, which required a substantially shorter growing season of 100 days, as opposed to traditional varieties, which needed 150 to 180 days. Due to this, multiple cropping became more prevalent in Asian countries.

According to the Government-owned North Korean news outlet KCNA, the late Kim Il Sung had a "profound knowledge of double cropping". Kim allegedly "tested double cropping in the experimental fields belonging to his residence for years in his lifetime".

One kind of multiple cropping is intercropping, where an additional crop is planted in the spaces available between the main crop.

Forest Gardening

Forest gardening is a food production and agroforestry system based on woodland ecosystems, incorporating fruit and nut trees, shrubs, herbs, vines and perennial vegetables which have yields directly useful to humans. Making use of companion planting, these can be intermixed to grow in a succession of layers, to replicate a woodland habitat.

Origins and History

Forest gardens are probably the world's oldest and most resilient agroecosystem. They originated in prehistoric times along jungle-clad river banks and in the wet foothills of monsoon regions. In the gradual process of families improving their immediate environment, useful tree and vine species were identified, protected and improved whilst undesirable species were eliminated. Eventually superior foreign species were selected and incorporated into the gardens.

Forest gardens are still common in the tropics and known by various names such as: *home gardens* in Kerala in South India, Nepal, Zambia, Zimbabwe and Tanzania; *Kandyan forest gardens* in Sri Lanka; *huertos familiares*, the "family orchards" of Mexico; and *pekarangan*, the gardens of "complete design", in Java. Forest gardens have been shown to be a significant source of income and food security for local populations.

Robert Hart

Robert Hart adapted forest gardening for temperate zones during the early 1960s. Hart began farming at Wenlock Edge in Shropshire with the intention of providing a healthy and therapeutic environment for himself and his brother Lacon. Starting as relatively conventional smallholders, Hart soon discovered that maintaining large annual vegetable beds, rearing livestock and taking care of an orchard were tasks beyond their strength. However, a small bed of perennial vegetables and herbs he planted was looking after itself with little intervention.

Following Hart's adoption of a raw vegan diet for health and personal reasons, he replaced his farm animals with plants. The three main products from a forest garden are fruit, nuts and green leafy vegetables. He created a model forest garden from a 0.12 acre (500 m^2) orchard on his farm and intended naming his gardening method *ecological horticulture* or *ecocultivation*. Hart later dropped these terms once he became aware that *agroforestry* and *forest gardens* were already being used to describe similar systems in other parts of the world. He was inspired by the forest farming methods of Toyohiko Kagawa and James Sholto Douglas, and the productivity of the Keralan home gardens as Hart explains:

From the agroforestry point of view, perhaps the world's most advanced country is the Indian state of Kerala, which boasts no fewer than three and a half million forest gardens...As an example of the extraordinary intensivity of cultivation of some forest gardens, one plot of only 0.12 hectare (0.3 acre) was found by a study group to have twenty-three young coconut palms, twelve cloves, fifty-six bananas, and forty-nine pineapples, with thirty pepper vines trained up its trees. In addition, the small holder grew fodder for his house-cow.

Seven-layer System

Robert Hart pioneered a system based on the observation that the natural forest can be divided into distinct levels. He used intercropping to develop an existing small orchard of apples and pears into an edible polyculture landscape consisting of the following layers:

1. 'Canopy layer' consisting of the original mature fruit trees.
2. 'Low-tree layer' of smaller nut and fruit trees on dwarfing root stocks.
3. 'Shrub layer' of fruit bushes such as currants and berries.
4. 'Herbaceous layer' of perennial vegetables and herbs.
5. 'Ground cover layer' of edible plants that spread horizontally.
6. 'Rhizosphere' or 'underground' dimension of plants grown for their roots and tubers.
7. 'Vertical layer' of vines and climbers.

A key component of the seven-layer system was the plants he selected. Most of the traditional vegetable crops grown today, such as carrots, are sun loving plants not well selected for the more shady forest garden system. Hart favoured shade tolerant perennial vegetables.

Development

The Agroforestry Research Trust, managed by Martin Crawford, runs experimental forest gardening projects on a number of plots in Devon, United Kingdom.

Forest gardening has been adopted as a common permaculture design element. Bill Mollison, who coined the term permaculture, visited Robert Hart at his forest garden in Wenlock Edge in October 1990. Numerous permaculturists are proponents of forest gardens, or food forests, such as Patrick Whitefield, Dave Jacke, Eric Toensmeier and Geoff Lawton. Whitefield wrote the book *How to Make a Forest Garden* in 2002, Jacke and Toensmeier co-authored the two volume book set *Edible Forest Gardening* in 2005, and Lawton presented the film *Establishing a Food Forest* in 2008.

Ken Fern had the idea that for a successful temperate forest garden a wider range of edible shade tolerant plants would need to be used. To this end, Fern created the organisation Plants for a Future (PFAF) which compiled a plant database suitable for such a system. Fern used the term *woodland gardening*, rather than forest gardening, in his book *Plants for a Future.*

The Movement for Compassionate Living (MCL) promote forest gardening and other types of vegan organic gardening to meet society's needs for food and natural resources. Kathleen Jannaway, the founder of MCL, wrote a book outlining a sustainable vegan future called *Abundant Living in the Coming Age of the Tree* in 1991. In 2009, the MCL provided a grant of £1,000 to the Bangor Forest Garden project in Gwynedd, North West Wales.

Projects

Forest gardens are to be found in various research trials such as those at the Central Rocky Mountain Permaculture Institute, community farms and gardens like Montview Neighbourhood farm, and in small yards throughout the temperate world.

In Canada food forester Richard Walker has been developing and maintaining food forests in the province of British Columbia for over 30 years. He developed a 3-acre food forest that when at maturity provided raw materials for a nursery and herbalism business as well as food for his family.

No Dig Gardening

No-dig gardening is a non-cultivation method used by some organic gardeners. The origins of no-dig gardening are unclear, and may be based on pre-industrial or nineteenth-century farming techniques. Two pioneers of the method in the twentieth century included F. C. King, Head Gardener at Levens Hall, South Westmorland, in the Lake District of England, who posed the Question "Is Digging Necessary?" in 1946 and a mysterious gardener from Middlecliffe in the UK, A. Guest, who in 1948 published the book Gardening Without

Digging. The work of these gardeners was supported by the Good Gardeners Association in the UK. Masanobu Fukuoka started his pioneering research work in this domain in 1938, and began publishing in the 1970s his Fukuokan philosophy of "Do Nothing Farming", which is now acknowledged by some as the tap root of the Permaculture movement. No-dig gardening was also promoted by Australian Esther Deans in the 1970s, and American gardener Ruth Stout advocated a "permanent" garden mulching technique in *Gardening Without Work* and no-dig methods in the 1950s and 1960s.

This technique recognises that micro- and macro-biotic organisms constitute a "food web" community in the soil, necessary for the healthy cycling of nutrients and prevention of problematic organisms and diseases. The plants transfer a portion of the carbon energy they produce to the soil, and microbes that benefit from this energy in turn convert available organic substances in the soil to the mineral elements the plants need to thrive.

Pollination Management

Figure 1: Honey bee on domestic plum blossom (Honey bees are especially[citation needed] well adapted to collecting and moving pollen,thus are the most commonly used pollinators. Note the light brown pollen in the pollen basket)

Pollination Management is the label for horticultural practices that accomplish or enhance pollination of a crop, to improve yield or quality, by understanding of the particular crop's pollination needs, and by knowledgeable management of pollenizers, pollinators, and pollination conditions.

Pollinator Decline

With the decline of both wild and domestic pollinator populations, pollination management is becoming an increasingly important part of horticulture. Factors that cause the loss of pollinators include pesticide misuse, unprofitability of beekeeping for honey, rapid transfer of pests and diseases to new areas of the globe, urban/suburban development, changing crop patterns, clearcut logging (particularly when mixed forests are replaced by monoculture pine), clearing of hedgerows and other wild areas, loss of nectar corridors for migratory pollinators, and human paranoia of stinging insects (killer bee hype).

In 1989, following Hurricane Hugo, massive aerial applications for mosquitoes were done in South Carolina. The following year, watermelon growers who did not place beehives in the fields, observed the fruit begin to develop, then abort, or develop into small deformed fruit. There were entire fields that never yielded a single usable melon. Some growers went out of business; others began to seriously manage pollination. Since beekeepers were also heavily damaged by the mosquito spraying, the supply of bees for pollination was critically short for several years.

Importance of Pollination Management

The increasing size of fields and orchards (monoculture) increase the importance of pollination management. Monoculture can cause a brief period when pollinators have more food resources than they can use, while other periods of the year can bring starvation or pesticide contamination of food sources. Most pollinator species rely on a steady nectar source and pollen source throughout the growing season to build up their numbers.

Crops that traditionally have had managed pollination include apple, almonds, pears, some plum and cherry varieties, blueberries, cranberries, cucumbers, cantaloupe, watermelon, alfalfa seeds, onion seeds, and many others. Some crops that have traditionally depended entirely on chance pollination by wild pollinators need pollination management nowadays to make a profitable crop.

Some crops, especially when planted in a monoculture situation, require a very high level of pollinators to produce economically viable crops. This may be because of lack of attractiveness of the blossoms, or from trying to pollinate with an alternative when the native pollinator is extinct or rare. These include crops such as alfalfa, cranberries, and kiwifruit. This technique is known as saturation pollination. In many such cases, various native bees are vastly more efficient at pollination (e.g., with blueberries), but the inefficiency of the honey bees is compensated for by using large numbers of hives, the total number of foragers thereby far exceeding the local abundance of native pollinators. In a very few cases, it has been possible to develop commercially viable pollination techniques that use the more efficient pollinators, rather than continued reliance on honey bees, as in the management of the alfalfa leafcutter bee.

Table: Number of hives needed per unit area of crop pollination

Common name	*number of hivesper acre*	*number of hivesper hectare*	*number of bee visits per square metre/minute*
Alfalfa	1, (3–5)	2.5, (4.9–12)	
Almonds	2–3	4.9–7.4	
Apples (normal size)	1	2.5	
Apples (semi dwarf)	2	4.9	
Apples (dwarf)	3	7.4	
Apricots	1	2.5	
Blueberries	3–4	7.4–9.9	2.5
Borage	0.6–1.0	1.5–2.5	
Buckwheat	0.5–1	1.2–2.5	
Canola	1	2.5	
Canola (hybrid)	2.0–2.5	4.9–6.2	
Cantaloupes	2–4, (average 2.4)	4.9–9.9, (average 5.9)	
Clovers	1–2	2.5–4.9	
Cranberries	3	7.4	
Cucumbers	1–2, (average 2.1)	2.5–4.9, (average 5.2)	
Ginseng	1	2.5	
Muskmelon	1–3	2.5–7.4	
Nectarines	1	2.5	
Peaches	1	2.5	
Pears	1	2.5	
Plums	1	2.5	
Pumpkins	1	2.5	
Raspberries	0.7–1.3	1.7–3.2	
Squash	1–3	2.5–7.4	
Strawberries	1–3.5	2.5–8.6	
Sunflower	1	2.5	
Trefoil	0.6–1.5	1.5–3.7	
Watermelon	1–3, (average 1.3)	2.5–4.9, (average 3.2)	
Zucchini	1	2.5	

It is estimated that about one hive per acre will sufficiently pollinate watermelons. In the 1950s when the woods were full of wild bee trees, and beehives were normally kept on most South Carolina farms, a farmer who grew ten acres (4 ha) of watermelons would be a large grower and probably had all the pollination needed. But today's grower may grow 200 acres (80 ha), and, if lucky, there might be one bee tree left within range. The only option in the current economy is to bring beehives to the field during blossom time.

Types of Pollinators

Organisms that are currently being used as pollinators in managed pollination are honey bees, bumblebees, alfalfa leafcutter bees, and orchard mason bees. Other species are expected to be added to this list as this field develops. Humans also can be pollinators, as the gardener who hand pollinates her squash blossoms, or the Middle Eastern farmer, who climbs his date palms to pollinate them.

The Cooperative extension service recommends one honey bee hive per acre (2.5 hives per hectare) for standard watermelon varieties to meet this crop's pollination needs. In the past, when fields were small, pollination was accomplished by a mix of bees kept on farms, bumblebees, carpenter bees, feral honey bees in hollow trees and other insects. Today, with melons planted in large tracts, the grower may no longer have hives on the farm; he may have poisoned many of the pollinators by spraying blooming cotton; he may have logged off the woods, removing hollow trees that provided homes for bees, and pushed out the hedgerows that were home for solitary native bees and other pollinating insects.

Planning for Improved Pollination

Before pollination needs were understood, orchardists often planted entire blocks of apples of a single variety. Because apples are self sterile, and different members of a single variety are genetic clones (equivalent to a single plant), this is not a good idea. Growers now supply pollenizers, by planting crab apples interspersed in the rows, or by grafting crab apple limbs on some trees. Pollenizers can also be supplied by putting drum bouquets of crab apples or a compatible apple variety in the orchard blocks.

The field of pollination management cannot be placed wholly within any other field, because it bridges several fields. It draws from horticulture, apiculture, zoology (especially entomology), ecology, and botany.

Sheet Mulching

In permaculture, sheet mulching is an agricultural no-dig gardening technique that attempts to mimic natural forests' processes. When deployed properly and in combination with other permacultural principles, it can generate healthy, productive and low maintenance ecosystems.

Technique

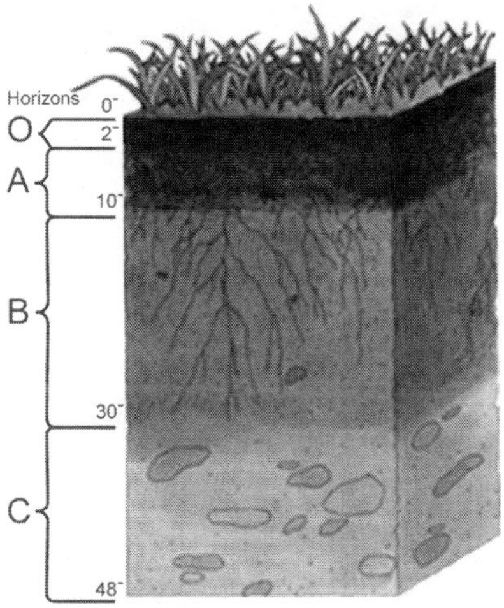

Figure 2: Typical layers of natural soil.

A model for sheet mulching consists of the following steps:

1. The area of interest is flattened by trimming down existing plant species such as grasses.
2. The soil is analysed and its pH is adjusted (if needed).
3. The soil is moisturized (if needed) to facilitate the activity of decomposers.
4. The soil is then covered with a thin layer of slowly-decomposing material (known as the weed barrier), typically cardboard. This suppresses the weeds by blocking sunlight, adds nutrients to the soil as weed matter quickly decays beneath the barrier, and increases the mechanical stability of the growing medium.
5. A layer (around 10 cm thick) of weed-free soil rich in nutrients is added, in an attempt to mimic the A horizon.
6. A layer (at most 15 cm thick) of weed-free, woody and leafy matter is added in an attempt to mimic the forest floor or O Horizon. Theoretically, the soil is now ready to receive the desirable plant seeds.

Variations and considerations

- Often the barrier is applied a few months before plantation to ensure the penetration of roots of newly planted seeds.
- Very thick barriers can cause anaerobic conditions.
- Some permaculturists incorporate composting in steps 5 and/ or 6.
- Sheets of newspaper and clothing can be used instead of cardboard.
- Before step 4, an initial layer (2-3 kg/m^2) of matter rich in nutrients (such as compost or manure) may be added to bolster decomposition.
- Some varieties of grasses and weeds may be beneficial in a number of ways. Such plants can be controlled and used rather than eradicated.
- One variation of sheet mulching, called hugelkultur, involves using buried logs and branches as the first layer of the bed.

Advantages

Sheet mulch has important advantages relative to conventional methods:

- Improvement of desirable plants' health and productivity.
- Retention of water and nutrients and stabilisation of biochemical cycles.
- Improvement of soil structure, soil life and prevention of soil erosion
- Avoidance of potentially dangerous pesticides, especially herbicides.
- Reduction of overall maintenance labour and costs.

Disadvantages

- Some weed seeds (such as those of Bermuda grass and species of bindweed) may persist under the barrier and within the soil seed bank.
- Slug populations may increase during the early stages of decomposition. However they can be kept away or harvested.
- The system may need a constant supply of organic material, at least during the early stages.
- Roaming animals may destroy the array.

Square Foot Gardening

Square Foot Gardening is the practice of planning small but intensively planted gardens. The phrase "square foot gardening" was popularised by Mel Bartholomew in a 1981 Rodale Press book and subsequent PBS television series. A full-length companion DVD, Square Foot Gardening (2010), was recently released in collaboration with Patti Moreno the Garden Girl. The practice combines concepts from other organic gardening methods, including a strong focus on compost, closely planted raised beds and biointensive attention to a small, clearly defined area. Proponents claim that the method is particularly well-suited for areas with poor soil, beginning gardeners or as adaptive recreation for those with disabilities.

Basic Method

The original square-foot-gardening method used an open-bottomed box to contain a finite amount of soil, which was divided with a grid into sections. To encourage variety of different crops over time, each square would be planted with a different kind of plant, the number of plants per square depending on an individual plant's size. A single tomato plant might take a full square, as might herbs such as oregano, basil or mint, while most strawberry plants could be planted four per square, with up to sixteen radishes per square. Tall or climbing plants such as maize or pole beans might be planted in a northern row (south in the southern hemisphere) so as not to shade other plants, and supported with lattice or netting.

The logic behind using smaller beds is that they are easily adapted, and the gardener can easily reach the entire area, without stepping on and compacting the soil. In the second edition, Bartholomew suggests using a "weed barrier" beneath the box, and filling it completely with "Mel's mix," a combination by volume of one third of decayed Sphagnum "peat moss", one-third expanded vermiculite and one-third blended compost. For accessibility, raised boxes may have bottoms to sit like tables at a convenient height, with approximately 6" (15cm) of manufactured soil per square foot.

Rationale

In this method, the garden space is divided into beds that are easily accessed from every side. A 4' x 4', 16 sq ft (1.5 m^2) or 120 cm x 120 cm, 1.4 m^2 garden is recommended for the first garden, and a path wide enough to comfortably work from should be made on each side of the bed, if possible, or if the bed must be accessed by reaching across it, a narrower one should be used so that no discomfort results from tending the garden. Each of the beds is divided into approximately one square foot units and marked out with sticks, twine, or sturdy slats to ensure that the square foot units remain visible as the garden matures.

Different seeds are planted in each square, to ensure a rational amount of each type of crop is grown, and to conserve seeds instead of overplanting, crowding and thinning plants. Common spacing is one plant per square for larger plants (broccoli, basil, etc.), four plants per square for medium large plants like lettuce, nine plants per square for medium-small plants like spinach, and sixteen per square for small plants such as onions and carrots. Plants that normally take up yards of space as runners, such as squash or cucumbers, are grown vertically on sturdy frames that are hung with netting or string to support the developing crops. Ones that grow deep underground, such as potatoes or carrots, are grown in a square foot section that has foot tall sides and a planting surface above the ground, so that a foot or more of framed soil depth is provided above the garden surface rather than below it.

The beds are weeded and watered from the pathways, so the garden soil is never stepped on or compacted. Because a new soil mixture is used to create the garden, and a few handfuls of compost are added with each harvest to maintain soil fertility over time, the state of the site's underlying soil is irrelevant. This gardening method has been employed successfully in every region, including in deserts, on high arid mountain plateaus, in cramped urban locations, and in areas with polluted or high salinity soils. It is equally useful for growing flowers, vegetables, herbs and some fruits in containers, raised beds, on tabletops or at ground level, in only 4 to 6 inches (150 mm) of soil. A few seeds per square foot, the ability to make compost, to water by hand, and to set up the initial garden in a sunny position or where a container, table or platform garden may be moved on wheels to receive light is all that is needed to set up a square foot garden.

Benefits of Square Foot Gardening

- Much less work. Conventional gardening requires heavy tools to loosen the soil, whereas in this method, the soil is never compacted and it remains loose and loamy. Weeding takes only seconds to minutes, due to the light soil, raised beds, and easily accessed plants. Harvests per foot of garden are increased due to the rich soil mixture, well-spaced plants, and lack of weeds produced when following Mel Bartholomew's method.

- Water Savings. The soil mixture that is advised has water-holding capacities, so that the garden needs water less frequently, and in much smaller quantities than when using other gardening methods. Water is also spared by hand-watering directly at the plant roots, so that there is very little waste
- Very little weeding. One benefit of this close planting is that the vegetables form a living mulch, and shade out many weed seeds before they have a chance to germinate.
- Pesticide / Herbicide Free. Natural insect repellent methods like companion planting (i.e. planting marigolds or other naturally pest-repelling plants) become very efficient in a close space and thus, pesticides are not necessary. The large variety of crops in a small space also prevents plant diseases from spreading easily
- Accessibility. A plywood bottom can be attached to the bottom of a box, which can then be placed on a tabletop or raised platform for those who wish to garden without bending or squatting, or to make gardening easy for wheelchair, cane or walker users.
- Covers and Cages. Because all beds are small (4'x4' or smaller), making covers or cages to protect plants from pests, cold, or sun is more practical than with larger gardens.

Succession Planting

In agriculture, succession planting refers to several planting methods that increase crop availability during a growing season by making efficient use of space and timing.

There are four basic approaches, that can also be combined:

- Two or more crops in succession: After one crop is harvested, another is planted in the same space. The length of the growing season, climate, and crop selection are key factors. For example, a cool season spring crop could be followed by a heat-loving summer crop.
- Same crop, successive plantings: Several smaller plantings are made at timed intervals, rather than all at once. The plants

mature at staggered dates, establishing a continuous harvest over an extended period. Lettuce and other salad greens are common crops for this approach.

- Two or more crops simultaneously: Non-competing crops, often with different maturity dates, are planted together in various patterns. Intercropping is one pattern approach; companion planting is a related, complementary practice.
- Same crop, different maturity dates: Several varieties are selected, with different maturity dates: early, main season, late. Planted at the same time, the varieties mature one after the other over the season.

These techniques can be used to design complex, highly productive cropping systems. The more involved the plan, the more detailed knowledge is required, of the specific varieties and how they perform in a particular growing location.

The term "succession planting" usually appears in literature for home gardening and small-scale farming, although the techniques apply to any scale. Some definitions include one or more, but not all of the four techniques described above.

Succession planting is often used in organic farming. Multiple cropping describes essentially the same general method. A catch crop refers to a specific type of succession planting, where a fast-growing crop is grown simultaneously with, or between successive plantings of, a main crop.

Weed Control

Weed control is the botanical component of pest control, using physical and chemical methods to stop weeds from reaching a mature stage of growth when they could be harmful to domesticated plants and livestock. In order to reduce weed growth, many "weed control" strategies have been developed in order to contain the growth and spread of weeds. The most basic is ploughing which cuts the roots of annual weeds. Today, chemical weed killers known as herbicides are widely used.

The Effects of Weeds on Other Plants

Weeds can compete with productive crops or pasture, or convert productive land into unusable scrub. Weeds are also often poisonous, distasteful, produce burrs, thorns or other damaging body parts or otherwise interfere with the use and management of desirable plants by contaminating harvests or excluding livestock.

Weeds tend to thrive at the expense of the more refined edible or ornamental crops. They provide competition for space, nutrients, water and light, although how seriously they will affect a crop depends on a number of factors. Some crops have greater resistance than others- smaller, slower growing seedlings are more likely to be overwhelmed than those that are larger and more vigorous. Onions are one of the crops most susceptible to competition, for they are slow to germinate and produce slender, upright stems. Quick growing, broad leafed weeds therefore have a distinct advantage, and if not removed, the crop is likely to be lost. Broad beans however produce large seedlings, and will suffer far less profound effects of weed competition other than during periods of water shortage at the crucial time when the pods are filling out. Transplanted crops raised in sterile seed or potting compost will have a head start over germinating weed seeds.

Weeds also differ in their competitive abilities, and can vary according to conditions and the time of year. Tall growing vigorous weeds such as fat hen (*Chenopodium album*) can have the most pronounced effects on adjacent crops, although seedlings of fat hen that appear in late summer will only produce small plants. Chickweed (*Stellaria media*), a low growing plant, can happily co-exist with a tall crop during the summer, but plants that have overwintered will grow rapidly in early spring and may swamp crops such as onions or spring greens.

The presence of weeds does not necessarily mean that they are competing with a crop, especially during the early stages of growth when each plant can find the resources it requires without interfering with the others. However, as the seedlings' size increases, their root systems will spread as they each begin to require greater amounts of water and nutrients. Estimates suggest that weed and crop can co-exist harmoniously for around three weeks, therefore it is important that weeds be removed early on in order to prevent competition occurring. Weed competition can have quite dramatic effects on crop growth. Harold A Roberts cites research carried out with onions wherein *"Weeds were carefully removed from separate plots at different times during the growth of the crop and the plots were then kept clean. It was found that after competition had started, the final yield of bulbs was being reduced at a rate equivalent to almost 4% per day. So that by delaying weeding for another fortnight, the yield was cut to less than half that produced on ground kept clean all the time."* (*The Complete Know And Grow Vegetables*, Bleasdale, Salter and others, OUP 1991).

He goes on to record that *"by early June, the weight of weeds per unit area was twenty times that of the crop, and the weeds had already taken from the soil about half of the nitrogen and a third of the potash which had been applied"*.

Perennial weeds with bulbils, such as lesser celandine and oxalis, or with persistent underground stems such as couch grass (*Agropyron repens*) or creeping buttercup (*Ranunculus repens*) are able to store reserves of food, and are thus able to grow faster and with more vigour than their annual counterparts. There is also evidence that the roots of some perennials such as couch grass exude allelopathic chemicals which inhibit the growth of other nearby plants.

Weeds can also host pests and diseases that can spread to cultivated crops. Charlock and Shepherd's purse may carry clubroot, eelworm can be harboured by chickweed, fat hen and shepherd's purse, while the cucumber mosaic virus, which can devastate the cucurbit family, is carried by a range of different weeds including chickweed and groundsel.

However, at times the role of weeds in this respect can be over-rated. As far as insect pests are concerned, often the species that live on weeds are not the same as those that attack vegetable crops; *"Tests with the common cruciferous weeds such as shepherds purse have shown that they do not act as hosts for the larvae of the cabbage root fly. One exception was found to be the wild radish, but this is not usually a weed of established vegetable gardens"* (Roberts, *The Complete Know And Grow Vegetables*). However pests such as cutworms may first attack weeds then move on to cultivated crops.

While charlock, a common weed in southeastern USA, may be considered a weed by row crop growers, it is highly valued by beekeepers, who seek out places where it blooms all winter, thus providing pollen for honeybees and other pollinators. Its bloom is resistant to all but a very hard freeze, and even that will only kill it back briefly. By feeding an array of pollinators during a seasonal dearth, it can redound to the farmer's advantage. Many weeds are likewise highly beneficial to pollinators.

Methods

In domestic gardens, methods of weed control include covering an area of ground with several layers of wet newspaper or one black plastic sheet for several weeks. In the case of using wet newspaper, the multiple layers prevent light from reaching all plants beneath, which kills them. Saturating the newspaper with water daily speeds the decomposition of the dead plants. Any weed seeds that start to sprout because of the water will also be deprived of sunlight, be killed, and decompose. After several weeks, all germinating weed seeds present in the ground should be dead. Then the newspaper can be removed and the ground can be planted. The decomposed plants will help fertilize the plants or seeds planted later.

Figure 3: Pesticide-free thermic weed control with a weed burner on a potato field in Dithmarschen

Figure 4: Weeds are removed manually in large parts of India.

In the case of using the black plastic sheet, the greenhouse effect is used to kill the plants beneath the sheet. A 5-10 cm layer of wood chip mulch on the ground will also prevent most weeds from sprouting. Also, gravel can be spread over the ground as an inorganic mulch. In agriculture, irrigation is sometimes used as a weed control measure such as in the case of paddy fields. Many people find that although the black plastic sheeting is extremely effective at preventing the weeds in areas where it covers, but in actual use it is difficult to achieve full coverage.

Knowing how weeds reproduce, spread and survive adverse conditions can help in developing effective control and management strategies. Weeds have a range of techniques that enable them to thrive;

Annual and biennial weeds such as chickweed, annual meadow grass, shepherd's purse, groundsel, fat hen, cleaver, speedwell and hairy bittercress propagate themselves by seeding. Many produce huge numbers of seed several times a season, some all year round. Groundsel can produce 1000 seed, and can continue right through a mild winter, whilst Scentless Mayweed produces over 30,000 seeds per plant. Not all of these will germinate at once, but over several seasons, lying dormant in the soil sometimes for years until exposed to light. Poppy seed can survive 80-100 years, dock 50 or more. There can be many thousands of seeds in a square foot or square metre of ground, thus and soil disturbance will produce a flush of fresh weed seedlings.

"Stale Seed Bed" Technique

One technique employed by growers is the 'stale seed bed', which involves cultivating the soil, then leaving it for a week or so.

When the initial flush of weeds has germinated, the grower will lightly hoe off before the desired crop is planted. However, even a freshly cleared bed will be susceptible to airborne seed from elsewhere, as well as seed brought in by passing animals which can carry them on their fur, or from freshly imported manure. The organic solution to the problem of spreading annual weeds lies in regular, properly timed weeding, preferably just before flowering (fortuitously, this is also the time at which they will be of the most value in composting).

This technique is also quite often used by farmers who let weeds germinate then return the soil before crop sowing.

Perennial weeds also propagate by seeding; the airborne seed of the dandelion and the rose-bay willow herb are parachuted far and wide. But they also have an additional range of vegetative means of spreading that gives them their pernicious reputation. Dandelion and dock put down deep tap roots, which, although they do not spread underground, are able to regrow from any remaining piece left in the ground. Removal of the complete tap root is the only sure remedy.

The most persistent of the perennials are those that spread by underground creeping rhizomes that can regrow from the tiniest fragment. These include couch grass, bindweed, ground elder, nettles, rosebay willow herb, Japanese knotweed, horsetail and bracken, as well as creeping thistle, whose tap roots can put out lateral roots. Other perennials put out runners that spread along the soil surface. As they creep along they set down roots, enabling them to colonise bare ground with great rapidity. These include creeping buttercup and ground ivy. Yet another group of perennials propagate by stolons-stems that arch back into the ground to reroot. Most familiar of these is the bramble.

All of the above weeds can be very difficult to eradicate- thick black plastic mulches can be effective to a degree, although will probably need to be left in place for at least two seasons. In addition, hoeing off weed leaves and stems as soon as they appear can eventually weaken and kill the plants, although this will require persistence in the case of plants such as bindweed. Nettle infestations can be tackled by cutting back at least three times a year, repeated over a three year period. Bramble can be dealt with in a similar way. Some plants are said to produce root exudates that suppress herbaceous weeds. *Tagetes minuata* is claimed to be effective against couch and ground elder, whilst a border of comfrey is also said to act as a barrier against the invasion of some weeds including couch.

Use of Herbicides

The above described methods of weed control avoid using chemicals. They are often used by farmers. However, these methods may damage a fragile soil by restructuring it, hence are not always used. They are those preferred by the organic gardener or organic farmer.

However weed control can also be achieved by the use of herbicides. Selective herbicides kill certain targets while leaving the desired crop relatively unharmed. Some of these act by interfering with the growth of the weed and are often based on plant hormones. Herbicides are generally classified as follows;

- Contact herbicides destroy only that plant tissue in contact with the chemical spray. Generally, these are the fastest acting herbicides. They are ineffective on perennial plants that are able to regrow from roots or tubers.
- Systemic herbicides are foliar-applied and are translocated through the plant and destroy a greater amount of the plant tissue. Modern herbicides such as glyphosate are designed to leave no harmful residue in the soil.

- Soil-borne herbicides are applied to the soil and are taken up by the roots of the target plant.
- Pre-emergent herbicides are applied to the soil and prevent germination or early growth of weed seeds.

In agriculture large scale and systematic weeding is usually required, often by machines, such as liquid herbicide sprayers, or even by helicopter (such as in the USA), to eliminate the massive amount of weeds present on farming lands.

However there are a number of techniques that the organic farmer can employ such as mulching and carefully timed cutting of weeds before they are able to set seed.

Organic Methods

Figure 5: A mechanical weed control device: the diagonal weeder

Typically a combination of methods are used in organic situations.

- Drip irrigation: Rubber hoses and other methods are used to bring water directly to the roots of the desired plants. This limits weed access to water.
- Manually pulling weeds: Labourers are used to pull weeds at various points in the growing process.
- Mechanically tilling around plants: Tractors are used to carefully till weeds around the crop plants at various points in the growing process. Besides tilling, other mechanical weed control methods also exist
- Ploughing: Ploughing includes tilling of soil, intercultural ploughing and summer ploughing. Ploughing through tilling of soil uproots the weeds which causes them to die. In summer

ploughing is done during deep summers. Summer ploughing also helps in killing pests.

- Crop rotation: Rotating crops with ones that kill weeds by choking them out, such as hemp, *Mucuna pruriens*, and other crops, can be a very effective method of weed control. It is a way to avoid the use of herbicides, and to gain the benefits of crop rotation.
- Weed mat: A weed mat is an artificial mulch, fibrous cloth material, bark or newspaper laid on top of the soil preventing weeds from growing to the surface.

Thermal Methods

There are several thermal methods known to control weed. Weed burners heat up soil quickly and destroy superficial parts of the plants. Weed seeds are often heat resistant and even react with an increase of growth on dry heat. Since the 19th century soil steam sterilization is used as a farming technique to clean soil completely from weeds. Several research results confirm the high effectiveness of humid heat against weeds and its seeds.

Wood

Besides those kinds of weeds which are of an herbaceous nature, there are others which are woody, and grow to a very considerable size; such as broom, furze and Lantanas. The first may be destroyed by frequent ploughing and harrowing, in the same manner as other perennial weeds are. Another method of destroying broom is by pasturing the field where it grows with grazers.

The best method of extirpating furze is to set fire to it in frosty weather, for frost has the effect of withering and making them burn readily. The stumps must then be cut over with a hatchet, and when the ground is well softened by rain it may be ploughed up, and the roots taken out by a harrow adapted to that purpose. If the field is soon laid down to grass, they will again spring up; in this case, pasturing with grazers is an effectual remedy. The thorn, or bramble, can only be extirpated by ploughing up the ground and collecting the roots.

In June, weeds are in their most succulent state, and in this condition, after they have lain a few hours to wither, cattle will eat almost every species. There is scarcely a hedge, border, or a nook, but what at that season is valuable.

UK Legislation Concerning Weeds

The *Weeds Act, 1959* is described as *"Preventing the spread of harmful or injurious weeds"*, and is mainly relevant to farmers and other rural settings rather than the allotment or garden-scale grower. There are five 'injurious' (that is, likely to be harmful to agricultural production) weeds covered by the provisions of the Weeds Act. These are:

- Spear thistle (*Cirsium vulgare*)
- Creeping or field thistle (*Cirsium arvense*)
- Curled Dock (*Rumex crispus*)
- Broad leaved dock (*Rumex obtusifolius*)
- Common ragwort (*Senecio jacobaea*) (n.b., this weed is poisonous to livestock. Livestock should not be allowed to graze where ragwort has grown until it is eradicated, and any traces have disintegrated. Ragwort should not be allowed to be harvested in hay or silage for feed).

The Department for Environment, Food and Rural Affairs (DEFRA) provides guidance for the treatment removal of these weeds from infested land. Much of this is oriented towards the use of herbicides, the majority of which may not be acceptable to the organic producer (apart from non-synthetic substances like sulphur, which in some circumstances are accepted within Soil Association standards) but in most cases there are manual techniques that can be used such as digging out the roots, mulching out or carefully timed cutting before seeds are able to spread.

Primary responsibility for weeds control rests with the occupier of the land on which the weeds are growing, therefore it is important to be alert to potential weed problems and to take prompt action. However, it should be remembered that most common farmland weeds are not "injurious" within the meaning of the Weeds Act, and many such plant species have conservation and environmental value. When dealing with complaints under the Weeds Act, DEFRA has a duty in law to try to achieve a reasonable balance among different interests. These include agriculture, countryside conservation and the public in general. Constructive discussion about problems caused by weeds can often result in effective solutions and avoid the need for DEFRA to take official action. In addition to those weeds covered by the 1959 act, under section 14 of the *Wildlife and Countryside Act 1981*, it can be an offence to plant or grow certain specified plants in the wild, including Giant Hogweed and Japanese Knotweed. Problems involving these plants can be referred to the local authority for the area where those weeds are growing as some local authorities have bye-laws controlling these plants. There is no statutory requirement for landowners to remove these plants from their property.

Stale Seed Bed

A false or stale seed bed is a useful organic weed control technique which involves creating a seedbed some weeks before seed is due to be sown. Preparation of such seedbed makes sure that any weed seeds that have been disturbed and brought to the soil surface during cultivation will thus have a chance to germinate, and can then be hoed off before sowing of the actual crop is carried out.

The technique can be utilised in early spring, when the weather is still too cold for proper seed germination. Several passes are made with a rototiller or plow, then weed seeds are allowed to germinate as weather permits.

By tilling, the farmer increases the chance of weed seed germination by the same method as one would for favourable vegetable/ crops: the fine soil allows weed seed to grow rapidly by allowing the seed to open and the roots to spread easier than in compacted soil. Deep tilling will also bring dormant seed to the surface for germination. Some species of plant are known for seeds that can lay deeply buried in the soil for years before favourable conditions allow germination.

Timing is important, however. Weed seeds must be tilled/howed or otherwise destroyed before they themselves can create new seeds. By destroying them early, the farmer eliminates most of that season's annual weeds. Turning the dead weeds back into the soil also increases soil nutrient content, although this difference is slight.

In many cases, several tillings are done, perhaps every two weeks beginning in very early spring. This allows more and more weed seeds to germinate only to be killed off later. This eliminates more weeds, but care must be used to not delay planting of a desirable crop later than the crop needs for a successful season's growth.

After several years, most, if not all, weeds can be eliminated from the seed bank in the soil. In some cases the effect can be noticed in the same year the process is first carried out.

If the weed patch is vigorous enough, the farmer may mow the patch first before tilling. This allows for easier/quicker decomposition in the soil when the plants are turned under.

Some farmers are noted for applying a light and inexpensive fertilizer mix to the soil to cause even more weed seeds to germinate. Although the merits of this may seem a trifle at first, it can pay dividends in the long run by eliminating seeds earlier that otherwise would have sprouted in later years. This too is open to debate.

Xeriscaping (Water Wise Gardening)

Xeriscaping and xerogardening refers to landscaping and gardening in ways that reduce or eliminate the need for supplemental water from irrigation. It is promoted in regions that do not have easily accessible, plentiful, or reliable supplies of fresh water, and is gaining acceptance in other areas as climate patterns shift.

Term History

The word *xeriscaping* is a portmanteau of *xeros* îÞñïò (Greek for "dry") and *landscaping*, and *xeriscape* is used for this style of garden. *Xeriscaping* refers to a method of landscape design that minimises water use.

Many Denver Water employees helped coin the term xeriscape, and Xeriscape and the xeriscape logo are registered trademarks of Denver Water, the water department of Denver, Colorado.

In some areas, terms such as water-conserving landscapes, *drought-tolerant* landscaping, *zeroscaping,* and *smart scaping* are used instead. Plants whose natural requirements are appropriate to the local climate are emphasized, and care is taken to avoid losing water to evaporation and run-off. The specific plants used in xeriscaping depend upon the climate.

Xeriscape Advantages

- Lowers consumption of imported or ground water.
  - o More water available for other domestic and community uses and the environment.
- Less time and work needed for maintenance effort, with gardening simpler and less stressful.
  - o Little or no lawnmowing
- Xeriscape plants in appropriate planting design, and soil grading and mulching, takes full advantage of rainfall retention.
- When water restrictions are implemented, by municipality or water costs, xeriscape plants will tend to survive and thrive, while more ornamental plants may be unable to adapt.

Figure 6: Cacti are not the only low water consuming plants used in Xeriscaping.

Xeriscape at Botanic Gardens

The world's first Xeriscape Demonstration Garden (now named Dryland Mesa) was created at the Denver Botanic Gardens in 1986. Its design was based on the "Seven Principles" of Xeriscape.

The Sonora Desert Museum holds an annual Xeriscape contest. The Special Awards presented include The J.D. Di Meglio Artistry In Landscaping Award (unique beauty in Xeriscape landscaping) and The Community Xeriscape Leader Award ("a driving force in promoting Xeriscape principles and/or the use of native and arid-adapted plants").

X-Ratings

In 1982 the Denver water Department's Xeriscape Demonstration Garden was dedicated with an "X-rated party". The X stood for Xeriscape, but the humour was intended. Garden centres continued with a system of X-ratings, from X to XXX, to specify how much water individual types of plants need.

The Seven Principles of Xeriscape

The Xeriscape concept is based on The Seven Principles:

Plan and Design

Create a diagram, drawn to scale, that shows the major elements of your landscape, including house, driveway, sidewalk, deck or patio, existing trees and other elements.

Once you've completed a base plan of an existing site, think about how you want to use your new Xeriscape. Do you want it to be a place for dogs to run? Curb appeal? Frame or screen views? Create a conceptual plan (bubble diagram) that shows the areas for turf, perennial beds, views, screens, slopes, etc. Once finished, develop a planting plan that reinforces the areas in the appropriate scale.

Soil Amendment

Most plants will benefit from the use of compost, which will help the soil retain water. Some desert plants prefer gravel soils instead of well-amended soils. Plants should either fit the soil or soil should be amended to fit the plants.

Efficient Irrigation

Xeriscape can be irrigated efficiently by hand or with an automatic sprinkler system. Zone turf areas separately from other plants and use the irrigation method that waters the plants in each area most efficiently. For grass, use gear-driven rotors or rotary spray nozzles that have larger droplets and low angles to avoid wind drift. Spray, drip line or bubbler emitters are most efficient for watering trees, shrubs, flowers and groundcovers.

If you water by hand, avoid oscillating sprinklers and other sprinklers that throw water high in the air or release a fine mist. The most efficient sprinklers release big drops close to the ground.

Water deeply and infrequently to develop deep roots. Never water during the day to reduce water lost to evaporation. If you have an automatic sprinkling system, adjust your controller monthly to accommodate weather conditions. Also, install a rain sensor to shut off the device when it rains.

Appropriate Plant and Zone Selection

Different areas in your yard receive different amounts of light, wind and moisture. To minimise water waste, group together plants with similar light and water requirements, and place them in an area that matches these requirements. Put moderate-water-use plants in low-lying drainage areas, near downspouts, or in the shade of other plants. Your turf will require the most water and shrub/perennial beds will require approximately half the amount of water. Dry, sunny areas support low-water-use plants that grow well in our climate. Planting a variety of plants with different heights, colour and textures creates interest and beauty.

Mulch

Mulch keeps plant roots cool, prevents soil from crusting, minimises evaporation and reduces weed growth. Organic mulches, such as bark chips, pole peelings or wood grindings, should be applied 2 to 4 inches deep. Fibre mulches create a web that is more resistant to wind and rain washout. Inorganic mulches, such as rocks and gravel, should be applied 2 to 3 inches deep. Surrounding plants with rock makes the area hotter; limit this practice.

Alternative Turf

Native grasses (warm-season) that have been cultivated for turf lawns, such as buffalo grass and blue grama, can survive with a quarter of the water that bluegrass varieties need. Warm-season grasses are greenest in June through September and straw brown the rest of the year.

Native grasses (cool season) such as bluegrass and tall fescue, are greenest in the spring and fall and go dormant in the high heat of the summer. New cultivars of bluegrass, such as Reveille, and tall fescue, can reduce typical bluegrass water requirements by at least 30 percent. Fine fescues can provide substantial water savings and is best used in areas that receive low traffic or are in shady locations.

Use the appropriate grass and limit the amount of grass to reduce the watering and maintenance requirements.

Maintenance

All landscapes require some degree of care during the year. Turf requires spring and fall aeration along with regular fertilization every 6 to 8 weeks. Keep your grass height at 3 inches and allow the clippings to fall. Trees, shrubs and perennials will need occasional pruning to remove dead stems, promote blooming or control height and spread. Much of the removed plant material can be shredded and used in composting piles.

Greywater Irrigation

Greywater is wastewater generated from domestic activities such as laundry, dishwashing, and bathing, which can be recycled on-site for uses such as landscape irrigation and constructed wetlands. Greywater differs from water from the toilets which is designated sewage or blackwater to indicate it contains human waste.

Definition

Greywater gets its name from its cloudy appearance and from its status as being between fresh, potable water (known as "white water") and sewage water ("black water"). In a household context, greywater is the leftover water from baths, showers, hand basins and washing machines only. Some definitions of greywater include water from the kitchen sink. Any water containing human waste is considered black water.

Elimination of Greywater

Domestic wastewater is usually combined at the sewer, so that grey- and blackwaters are removed together using a shared sewerage system in a process called *elimination.*

Sewage water can then be treated to limit pollution and health risks, before being returned to the environment at large. Most greywater ends up as effluent in rivers and oceans in this way.

There are other alternatives to eliminating greywater that allow for efficient use; using it to irrigate plants is a common practice. The plants use contaminants of greywater, such as food particles, as nutrients in their growth. However, salt and soap residues can be toxic to microbial and plant life alike, but can be absorbed and degraded through constructed wetlands and aquatic plants such as sedges, rushes, and grasses.

Recycling

Most greywater is easier to treat and recycle than blackwater, because of lower levels of contaminants. While all greywater contains micro-organisms, the health hazards associated with greywater from a multiple-dwelling source should be considered differently from that of a single dwelling greywater source.

If collected using a separate plumbing system from blackwater, domestic greywater can be recycled directly within the home, garden or company and used either immediately or processed and stored. If stored, it must be used within a very short time or it will begin to putrefy due to the organic solids in the water. Recycled greywater of this kind is never safe to drink, but a number of stages of filtration and microbial digestion can be used to provide water for washing or flushing toilets. Some greywater may be applied directly from the sink to the garden or container field, receiving further treatment from soil life and plant roots. Given that greywater may contain nutrients, pathogens, and is often discharged warm, it is very important to store it before use in irrigation purposes, unless it is properly treated first.

Systems

At present, several water recycling systems exist which can be used to:

- recycle the water without purifying it
- recycle the water while purifying or decontaminating it

Water Recycling Systems Without Purification

Water recycling without purification is used in certain agricultural companies (e.g., tree nurseries) and dwellings for applications where potable water is not required (e.g., garden and land irrigation, toilet flushing). It may also be used in dwellings when the greywater (e.g., from rainwater) is already fairly clean to begin with and/or has not been polluted with non-degradable chemicals such as non-natural soaps (thus using natural cleaning products instead). This water system also needs a supply of water to recycle and reuses water as well. It is also not recommended to use water that has been in the greywater filtration system for more than 24 hours or bacteria builds up affecting the water that is being reused. Water purification/ decontamination systems then again are used for applications where potable water *is* required (e.g., to allow drinking, and/or for other domestic tasks as washing, showering).

Water Recycling with Purification

For filtering the water to become potable (or near-potable), there are numerous systems based on *soft* processes. These include natural biological principles such as

- mechanical systems (sand filtration, lava filter systems and systems based on UV radiation)
- biological systems (plant systems as treatment ponds, constructed wetlands, living walls) and Bio reactors or compact systems as activated sludge systems, biorotors, aerobic and anaerobic biofilters, submerged aerated filters, biorolls

Finally, "hard", direct processes, such as distillation (evaporation) or mechanical processes such as membrane filtration, (typically ultrafiltration and reverse osmosis, which are capable of treating high volumes of grey water) can create potable, or near-potable water. There seem to be no commercially available "hard" greywater recovery devices suitable for on-site use in the individual household, even though a number of such technologies exist.

In order to purify the potable water adequately, several of these systems are usually combined to work as a whole. Combination of the systems is done in two to three stages, using a primary and a secondary purification. Sometimes a tertiary purification is also added.

Some municipal sewage systems recycle a certain amount of grey and blackwater using a high standard of treatment, providing reclaimed water for irrigation and other uses.

Application of Recycled Greywater

Irrigation: Greywater typically breaks down faster than blackwater and has lower levels of nitrogen and phosphorus. However, all greywater must be assumed to have some blackwater-type components, including pathogens of various sorts. Greywater should be applied below the surface where possible (e.g., via drip line on top of the soil, under mulch; or in mulch-filled trenches) and not sprayed, as there is a danger of inhaling the water as an aerosol.

Figure 7: Treated greywater drips into a bucket

Long-term research on greywater use on soil has not yet been done and it is possible that there may be negative impacts on soil productivity.

In any greywater system, it is essential to put nothing toxic down the drain—no bleaches, bath salts, artificial dyes, cleansers, and no products containing boron (which is toxic to plants at high levels).

It is crucial to use all-natural, biodegradable soaps whose ingredients do not harm plants. Most powdered detergents, and some liquid detergents, are sodium-based, which can inhibit seed-germination and destroy the structure of clay soils.

Indoor Reuse

Recycled greywater from showers and bathtubs can be used for flushing toilets in most European and Australian jurisdictions and in United States jurisdictions that have adopted the International Plumbing Code.

Such a system could provide an estimated 30% reduction in water use for the average household. The danger of biological contamination is avoided by using:

- a cleaning tank, to eliminate floating and sinking items
- an intelligent control mechanism that flushes the collected water if it has been stored long enough to be hazardous; this completely avoids the problems of filtration and chemical treatment.

The Uniform Plumbing Code, adopted in some United States jurisdictions, prohibits greywater use indoors.

Extreme Living Conditions

Greywater use promotes the ability to build in areas unsuitable for conventional treatment, or where conventional treatment is costly. The Mars Desert Research Station uses greywater recycling, and might be used on trips to Mars to reduce water consumption and increase oxygen generation.

Heat Reclamation

Devices are currently available that capture heat from residential and industrial greywater, through a process called drainwater heat recovery, greywater heat recovery, or hot water heat recycling.

Rather than flowing directly into a water heating device, incoming cold water flows first through a heat exchanger where it is pre-warmed by heat from greywater flowing out from such activities as dishwashing, or showering. Typical household devices receiving greywater from a shower can recover up to 60% of the heat that would otherwise go to waste.

Ecology

Because greywater use, especially domestically, reduces demand on conventional water supplies and pressure on sewage treatment systems, its use is very beneficial to local waterways. In times of drought, especially in urban areas, greywater use in gardens or toilet systems helps to achieve some of the goals of ecologically sustainable development.

Benefits

The potential ecological benefits of greywater recycling include

- Lower fresh water extraction from rivers and aquifers
- Less impact from septic tank and treatment plant infrastructure
- Topsoil nutrification
- Reduced energy use and chemical pollution from treatment
- Groundwater recharge
- Increased plant growth
- Reclamation of nutrients
- Greater quality of surface and ground water when preserved by the natural purification in the top layers of soil than generated water treatment processes.

In the U.S. Southwest and the Middle East where available water supplies are limited, especially in view of a rapidly growing population, a strong imperative exists for adoption of alternative water technologies.

Governmental Regulation

Government regulation governing domestic greywater use for landscape irrigation (diversion for reuse) is still a developing area and continues to gain wider support as the actual risks and benefits are considered and put into clearer perspective.

'Greywater' (by pure legal definition) is considered in some jurisdictions to be 'sewage' (all wastewater including greywater and toilet waste), but in the U.S. states that adopt the *International Plumbing Code*, it can be used for sub surface irrigation and for toilet flushing, and in states that adopt the *Uniform Plumbing Code*, it can be used in underground disposal fields that are akin to shallow sewage disposal fields.

Wyoming allows surface and subsurface irrigation and other non specific use of greywater under a Department of Environmental Quality policy enacted in March, 2010. California, Utah, New Mexico and some other states allow true subsurface drip irrigation with greywater. Where greywater is still considered sewage, it is bound by the same regulatory procedures enacted to ensure properly engineered septic tank and effluent disposal systems are installed for long system life and to control spread of disease and pollution. In such regulatory jurisdictions, this has commonly meant domestic greywater diversion for landscape irrigation was either simply not permitted or was discouraged by expensive and complex sewage system approval requirements. Wider legitimate community greywater diversion for landscape irrigation has subsequently been handicapped and resulted in greywater reuse continuing to still be widely undertaken by householders outside of and in preference to the legal avenues.

However, with water conservation becoming a necessity in a growing number of jurisdictions, business, political and community pressure has made regulators seriously reconsider the actual risks against actual benefits.

It is now recognised and accepted by an increasing number of regulators that the microbiological risks of greywater reuse at the single dwelling level where inhabitants already had intimate knowledge of that greywater are in reality an insignificant risk, when properly managed without the need for complex, expensive and onerous red tape approval processes. This is reflected in the NSW Government Department of Water and Energy's newly released greywater diversion rules, and the recent passage of greywater legislation in Montana. In the 2009 Legislative Session, the state of Montana passed a bill expanding greywater use into multi-family and commercial buildings. The Department of Environmental Quality has already drafted rules and design guidelines for greywater reuse systems in all these applications. Existing staff would review systems proposed for new subdivisions in conjunction with review of all other wastewater system components.

Companion Plants

The purpose of this list is to provide plants that assist in the growth of others, be it attracting beneficial insects, naturally regulating pests, repelling harmful insects, providing nutrients, part of a biological pest control program, and in some cases simply a shaded microclimate or climbing support.

Neglected Crops

Neglected and underutilised crops are domesticated plant species that have been used for centuries or even millennia for their food, fibre, fodder, oil or medicinal properties, but have been reduced in importance over time owing to particular supply and use constraints. These can inter alia include poor shelf life, unrecognised nutritional value, poor consumer awareness and reputational problems ("poor people's food"). As the demand for plant and crop attributes changes (re-appraisal or discovery of nutritional traits, culinary value, adaptation to climate change, etc.), neglected crops can overcome the constraints to the wider production and use. As a matter of fact, many formerly neglected crops are now globally significant crops (oilpalm, soybean, kiwi fruit). Although the options for scaling-up neglected crops for large-scale agriculture appear to be increasingly exhausted, many species have the potential to contribute to food security, nutrition, dietary and culinary diversification, health and income generation. They also provide environmental services. It is impossible to define what would constitute "proper" or "correct" levels of utilisation, however, it is evident that many neglected species are underutilised relative to their nutritional value and productivity.

Just three crops - maize, wheat and rice - account for about 40% of the world's consumption of calories and protein. 95% of the world's food needs are provided for by just 30 species of plants. In stark contrast, at least 12,650 species names have been compiled as edible (Kunkel 1984).

Neglected and underutilised plants are those that could be - and, in many cases, historically have been - used for food and other uses on a larger scale. Such crop species have also been described as "minor", "orphan", "promising" and "little-used". They continue to play an important role in the subsistence and economy of poor people throughout the developing world, particularly in the agrobiodiversity-rich tropics. Despite their potential for dietary diversification and the provision of micro-nutrients such as vitamins and minerals, they continue to attract little research and development attention.

Alongside their commercial potential, many of the underutilised crops also provide important environmental services, as they are adapted to marginal soil and climate conditions.

It is difficult to precisely define which attributes makes a crop "underutilised", but often they display the following features:

- Linkage with the cultural heritage of their places of origin
- Local and traditional crops whose distribution, biology, cultivation and uses are poorly documented
- Adaptation to specific agro-ecological niches and marginal land
- Weak or no formal seed supply systems
- Traditional uses in localised areas
- Produced in traditional production systems with little or no external inputs

- Receive little attention from research, extension services, policy and decision makers, donors, technology providers and consumers
- May be highly nutritious and/or have medicinal properties or other multiple uses.

Organic Clothing

Organic clothing is clothing made from materials raised or grown in compliance with organic agricultural standards. Organic clothing uses cotton, jute, silk, ramie, or wool. Retailers charge more for organic clothing because such the material used are free from chemicals or genetically modified seeds.

Alternative terms are organic fashion, organic cloths.

Benefits

Authentic organic fabrics and clothing can help the environment in a number of ways, such as:

- Manufacture of chemicals is not required
- Chemical residues are not entered accidentally into the environment
- Humans and animals are not exposed to chemicals
- When the fabric is finished with chemicals are not returned to the earth in landfill, or enter into recycling process.

Cotton covers 2.5% of the world's cultivated land yet uses 16% of the world's insecticides, more than any other single major crop. It can take almost a 1/3 pound of synthetic fertilizers to grow one pound of raw cotton in the US, and it takes just under one pound of raw cotton to make one T-shirt.

Criticism

Many highstreet retailers market organic clothing ranges that contain chemicals from the dyeing to bleaching process, which is inconsistent with the idea of Organic Clothing. Many companies sell clothing made from bamboo, which is commonly labelled as "organic", however this is a false statement. Bamboo fabric is typically chemically manufactured by "cooking" the bamboo leaves and woody shoots in strong chemical solvents such as sodium hydroxide and carbon disulfide, in a process also known as hydrolysis alkalization combined with multi-phase bleaching. Both sodium hydroxide and carbon disulfide have been linked to serious health problems. This finished material is similar to rayon and modal, which are more accurate terms of describing bamboo fabrics. Criticism also concerns the high cost of the products.

Organic Food

Organic foods are foods that are produced using methods that do not involve modern synthetic inputs such as synthetic pesticides and chemical fertilizers, do not contain genetically modified organisms, and are not processed using irradiation, industrial solvents, or chemical food additives.

For the vast majority of human history, agriculture can be described as organic; only during the 20th century was a large supply of new synthetic chemicals introduced to the food supply. The organic farming movement arose in the 1940s in response to the industrialisation of agriculture known as the Green Revolution.

The weight of the available scientific evidence has not shown a significant difference between organic and more conventionally grown food in terms of safety, nutritional value, or taste.

Organic food production is a heavily regulated industry, distinct from private gardening. Currently, the European Union, the United States, Canada, Japan and many other countries require producers to obtain special certification in order to market food as organic within their borders. In the context of these regulations, organic food is food made in a way that complies with organic standards set by national governments and international organisations. In the United States, organic production is a system that is managed in accordance with the Organic Foods Production Act (OFPA) of 1990 and regulations in Title 7, Part 205 of the Code of Federal Regulations to respond to site-specific conditions by integrating cultural, biological, and mechanical practices that foster cycling of resources, promote ecological balance, and conserve biodiversity. If livestock are involved, the livestock must be reared with regular access to pasture and without the routine use of antibiotics or growth hormones. In the United States, a food can be labelled as "organic" if it contains a minimum of 95% organic ingredients. In most countries, organic produce may not be genetically modified. It has been suggested that the application of nanotechnology to food and agriculture is a further technology that needs to be excluded from certified organic food. The Soil Association (UK) has been the first organic certifier to implement a nano-exclusion.

Meaning and Origin of the Term

In 1939, Lord Northbourne coined the term *organic farming* in his book *Look to the Land* (1940), out of his conception of "the farm as organism," to describe a holistic, ecologically-balanced approach to farming—in contrast to what he called *chemical farming*, which relied on "imported fertility" and "cannot be self-sufficient nor an organic whole." This is different from the scientific use of the term "organic," to refer to a class of molecules that contain carbon, especially those involved in the chemistry of life. This class of molecules includes many chemicals that would not be used in organic farming.

Identifying Organic Food

Processed organic food usually contains only organic ingredients. If non-organic ingredients are present, at least a certain percentage of the food's total plant and animal ingredients must be organic (95% in the United States, Canada, and Australia) and any non-organically produced ingredients are subject to various agricultural requirements. Foods claiming to be organic must be free of artificial food additives, and are often processed with fewer artificial methods, materials and conditions, such as chemical ripening, food irradiation, and genetically modified ingredients. Pesticides are allowed so long as they are not synthetic.

Early consumers interested in organic food would look for non-chemically treated, non-use of unapproved pesticides, fresh or minimally processed food. They mostly had to buy directly from growers: "Know your farmer, know your food" was the motto. Personal definitions of what constituted "organic" were developed through firsthand experience: by talking to farmers, seeing farm conditions, and farming activities. Small farms grew vegetables (and raised livestock) using organic farming practices, with or without certification, and the individual consumer monitored. As demand for organic foods continued to increase, high volume sales through mass outlets such as supermarkets rapidly replaced the direct farmer connection. Today there is no limit to organic farm sizes and many large corporate farms currently have an organic division. However, for supermarket consumers, food production is not easily observable, and product labelling, like "certified organic", is relied on. Government regulations and third-party inspectors are looked to for assurance.

The USDA carries out routine inspections of farms that produce USDA Organic labelled foods. Of the 30 third party inspectors 15 of them have been placed under probation after an audit. On April 20,

2010, the Department of Agriculture said that it would begin enforcing rules requiring the spot testing of organically grown foods for traces of pesticides, after an auditor exposed major gaps in federal oversight of the organic food industry.

Legal Definition

To be certified organic, products must be grown and manufactured in a manner that adheres to standards set by the country they are sold in:

- Australia: Australian Organic Standard and NASAA Organic Standard
- Canada: Canada Gazette, Government of Canada
- European Union: EU-Eco-regulation
    - Sweden: KRAV
    - United Kingdom: Department for Environment, Food and Rural Affairs (DEFRA)
- Norway: Debio Organic certification
- India: NPOP, (National Program for Organic Production)
- Japan: JAS Standards
- United States: National Organic Program (NOP) Standards.

Environmental Impact

Several surveys and studies have attempted to examine and compare conventional and organic systems of farming. The general consensus across these surveys is that organic farming is less damaging for the following reasons:

- Organic farms do not consume or release synthetic pesticides into the environment—some of which have the potential to harm soil, water and local terrestrial and aquatic wildlife.
- Organic farms are better than conventional farms at sustaining diverse ecosystems, *i.e.*, populations of plants and insects, as well as animals.
- When calculated per unit area, organic farms use less energy and produce less waste, *e.g.*, waste such as packaging materials for chemicals.

A 2003 investigation by the Department for Environment Food and Rural Affairs in the UK found, similar to other reports, that organic farming "can produce positive environmental benefits", but that some of the benefits were decreased or lost when comparisons are made on "the basis of unit production rather than area".

The environmental impact of pesticides as well as the impact to the health of farm workers are reasons given for purchasing organic food. However, critics of organic farming methods believe that the increased land needed to farm organic food could potentially destroy the rainforests and wipe out many ecosystems.

Yield

One study found a 20% smaller yield from organic farms using 50% less fertilizer and 97% less pesticide. Studies comparing yields have had mixed results. Supporters claim that organically managed soil has a higher quality and higher water retention. This may help increase yields for organic farms in drought years.

One study from the Danish Environmental Protection Agency found that, area-for-area, organic farms of potatoes, sugar beet and seed grass produce as little as half the output of conventional farming. Michael Pollan, author of *The Omnivore's Dilemma*, responds to this by pointing out that the average yield of world agriculture is substantially lower than modern sustainable farming yields. Bringing average world yields up to modern organic levels could increase the world's food supply by 50%.

A 2007 study compiling research from 293 different comparisons into a single study to assess the overall efficiency of the two agricultural systems has concluded that:

> *...organic methods could produce enough food on a global per capita basis to sustain the current human population, and potentially an even larger population, without increasing the agricultural land base.*

The researchers also found that while in developed countries, organic systems on average produce 92% of the yield produced by conventional agriculture, organic systems produce 80% more than conventional farms in developing countries, because the materials needed for organic farming are more accessible than synthetic farming materials to farmers in some poor countries. On the other hand, communities that lack sufficient manure to replenish soils would struggle with organic farming, and the soil would degrade rapidly.

Energy Efficiency

A study of the sustainability of apple production systems showed that in comparing a conventional farming system to an organic method of farming, the organic system is more energy efficient. A more comprehensive study compared efficiency of agriculture for products such as grain, roughage crops, and animal husbandry. While the study did not investigate specific additional requirements of arable land or numbers of farm labourers to produce total yields for organic farming vs. conventional farming, leaving open the question of overall capacity of organic farming to meet current and future agricultural needs, it concluded that organic farming had a higher yield per unit of energy over multiple crops and for livestock. However, conventional farming had higher total yield. Conversely, another study noted that organic wheat and corn production was more energy efficient than conventional methods while organic apple and potato production was less energy efficient than conventional methods. A long-term study, spanning two decades, noted that crop yields were 20% lower in organic systems while fertilizer plus energy input was 34% to 53% lower. However, pesticide input was reduced by 97% in organic farm systems.

Consumer Safety

There is widespread belief that organic food is significantly safer for consumption than food grown conventionally, based mainly on anecdotal evidence and testimonials rather than scientific evidence, which has fuelled increased demand for organic food despite higher prices. Reviews of the available body of scientific literature comparing the safety of organic to conventional foods have found neither to be significantly more safe than the other. Firm conclusions about the relative safety of organic foods has been hampered by the difficulty in proper study design and relatively small number of studies directly comparing organic food to conventional food.

Claims of improved safety of organic food has largely focused on pesticide residues. While studies have shown organically grown fruits and vegetables have significantly lower pesticide residue levels, the significance of this finding on actual health risk reduction is debatable as both conventional foods and organic foods generally have pesticide levels well below government established guidelines for what is considered safe. This view has been echoed by the U.S. Department of Agriculture and the UK Food Standards Agency. Claims of increased risk related to pesticide residue and rates of infertility or lower sperm counts have not been supported by the evidence in the medical literature. Reviews have noted that the risks from microbiological sources or natural toxins are likely to be much more significant than short term or chronic risks from pesticide residues.

Some focus has been placed on the amount of nitrogen content in certain vegetables, especially green leafy vegetables and tubers when grown organically as compared to conventionally. While these vegetables, when grown organically, have been found to have lower nitrogen content, there is no consensus as to whether consumption of lower levels of nitrogen translates to improved health risk. When evaluating environmental toxins such as heavy metals, the USDA noted that organically raised chicken may have lower arsenic levels, while a literature review found no significant evidence that levels of arsenic, cadmium or other heavy metals differed significantly between organic and conventional food products.

In looking at possible increased risk to safety from organic food consumption, reviews have found that although there are theoretical increased risk from microbiological contamination due to increased manure use as fertilizer from organisms like *E. coli* O157:H7 during organic produce production, there does not exist sufficient evidence of actual incidence of outbreaks that can be clearly tied to organic food production to draw any firm conclusions. Other possible sources of increased safety risk from organic food consumption like use of biological pesticides or the theoretical risk from mycotoxins from fungi grown on products due to the lack of effective organic compliant fungicides have likewise not been confirmed by rigorous studies in the scientific literature.

Nutritional Value and Taste

According to the UK's Food Standards Agency, "Consumers may choose to buy organic fruit, vegetables and meat because they believe them to be more nutritious than other food. However, the balance of current scientific evidence does not support this view." A 12-month systematic review commissioned by the FSA in 2009 and conducted at the London School of Hygiene & Tropical Medicine based on 50 years' worth of collected evidence concluded that "there is no good evidence that consumption of organic food is beneficial to health in relation to nutrient content." Other studies have found no proof that organic food offers greater nutritional values, more consumer safety or any distinguishable difference in taste. A review of nutrition claims showed that organic food proponents are unreliable information sources which harm consumers, and that consumers are wasting their money if they buy organic food believing that it contains better nutrients.

Although it is commonly claimed that organically grown food tastes better than conventionally grown food, reviews of the literature that looked at the sensory qualities of the two have not found convincing evidence that there is any significant differences.

Organic Baby Products

Organic baby products are commercially produced infant formulas and baby foods that have obtained organic certification. They are often recommended as substitutes for conventionally marketed baby foods in order to reduce the amount of chemical exposure for rapidly developing infants. Since babies' bodies are smaller and their brains grow faster, they are more vulnerable to environmental toxins as they absorb more pesticides per pound of body weight than adults.

In many countries mothers who are not breastfeeding can find organic versions of infant formulas and baby foods using various substances (e.g. soy, dairy milk, wheat) through various channels (e.g. hospitals, grocery stores, mail order). However the marketing of organic baby products for newborns and infants under six months of age, just like the marketing of other breastmilk substitutes, has met with increasing skepticism and controversy. The World Health Organisation, UNICEF and many national health departments recommend that infants should be exclusively breastfed for the first six months of life to achieve optimal health, growth and development.

Organic baby products are more expensive to produce and get certified, therefore often priced higher than conventional goods and not as accessible to middle and lower income families. The International Federation of Organic Agriculture Movements (IFOAM) says that while prices have been declining, there are reasons the costs of certified organic products remain higher than conventional products. They include limited supply as compared to demand, greater labour and transportation costs per unit produced. The smaller number of units involved also means marketing and distribution costs have to be spread out over a smaller number of units. As demand for organic food and products increases, technological innovations and economies of scale should reduce costs of production, processing, distribution and marketing for organic produce.

Economics

Demand for organic foods is primarily concern for personal health and concern for the environment. Organic products typically cost 10 to 40% more than similar conventionally produced products. According to the USDA, Americans, on average, spent $1,347 on groceries in 2004; thus switching entirely to organics would raise their cost of groceries by about $135 to $539 per year ($11 to $45 per month) assuming that prices remained stable with increased demand. Processed organic foods vary in price when compared to their conventional counterparts.

While organic food accounts for 1–2% of total food sales worldwide, the organic food market is growing rapidly, far ahead of the rest of the food industry, in both developed and developing nations.

- World organic food sales jumped from US $23 billion in 2002 to $52 billion in 2008.
- The world organic market has been growing by 20% a year since the early 1990s, with future growth estimates ranging from 10%–50% annually depending on the country.

North America

United States:

- Organic food is the fastest growing sector of the American food marketplace.
- Organic food sales have grown by 17 to 20 percent a year for the past few years while sales of conventional food have grown at only about 2 to 3 percent a year.
- In 2003 organic products were available in nearly 20,000 natural food stores and 73% of conventional grocery stores.
- Organic products accounted for 3.7% of total food and beverage sales, and 11.4% of all fruit and vegetable sales in the year 2009.
- Two thirds of organic milk and cream and half of organic cheese and yogurt are sold through conventional supermarkets.

Canada:

- Organic food sales surpassed $1 billion in 2006, accounting for 0.9% of food sales in Canada.
- Organic food sales by grocery stores were 28% higher in 2006 than in 2005.
- British Columbians account for 13% of the Canadian population, but purchased 26% of the organic food sold in Canada in 2006.

Europe

In the European Union (EU25) 3.9% of the total utilised agricultural area was used for organic production in 2005. The countries with the highest proportion of organic land were Austria (11%) and Italy (8.4), followed by the Czech Republic and Greece (both 7.2%). The lowest figures were shown for Malta (0.1%), Poland (0.6%) and Ireland (0.8%).

Austria:

- 11.6% of all farmers produced organically in 2007. The government has created incentives to increase the figure to 20% by 2010.
- In 2006, 4.9% of all food products sold in Austrian supermarkets (including discount stores) were organic. In the same year, 8,000 different organic products were available.

Italy:

- Since 2000, the use of some organic food is compulsory in Italian schools and hospitals. A 2002 law of the Emilia Romagna region implemented in 2005, explicitly requires that the food in nursery and primary schools (from 3 months to 10 years) must be 100% organic, and the food in meals at schools, universities and hospitals must be at least 35% organic.

Poland:

- In 2005 168,000 ha of land were under organic management. 7 percent of Polish consumers buy food that was produced according to the EU-Eco-regulation. The value of the organic market is estimated at 50 million Euros (2006).

UK:

- Organic food sales increased from just over £100 million in 1993/94 to £1.21 billion in 2004 (an 11% increase on 2003).

Latin America

Cuba:

- After the collapse of the Soviet Union in 1991, agricultural inputs that had previously been purchased from Eastern bloc countries were no longer available in Cuba, and many Cuban farms converted to organic methods out of necessity. Consequently, organic agriculture is a mainstream practice in Cuba, while it remains an alternative practice in most other countries. Although some products called organic in Cuba would not satisfy certification requirements in other countries (crops may be genetically modified, for example), Cuba exports organic citrus and citrus juices to EU markets that meet EU organic standards. Cuba's forced conversion to organic methods may position the country to be a global supplier of organic products.

Mexico:

- Although certification seals have become more common on Mexican produce, mainly for export, there have been strong internal movements to block GMOs and decrease pesticide practice. As the country does not have as long a history of using pesticides as developed nations, it is striving to regain the purity of much of its arable land.
- Much of its internal and domestic produce consumption is not labeled as organic, even when the practice is sustainable. This is mainly due to cultural differences.

Organic Movement

The organic movement broadly refers to the organisations and individuals involved worldwide in the promotion of organic farming, which is a more sustainable mode of agriculture. Its history goes back to the first half of the 20th century, when modern large-scale agricultural practices began to appear.

Overview and Origin

The organic movement began in the early 1900s in response to the shift towards synthetic nitrogen fertilizers and pesticides in the early days of industrial agriculture. It lay dormant for many years, kept alive by a relatively small group of ecologically minded farmers. These farmers came together in various associations: Demeter International of Germany, which encouraged biodynamic farming and began the first certification program, the Australian Organic Farming and Gardening Society, the Soil Association of the United Kingdom, and Rodale Press in the United States, along with others. In 1972 these organisations joined to form the International Federation of Organic Agriculture Movements (IFOAM). In recent years, environmental awareness has driven demand and conversion to organic farming. Some governments, including the European Union, have begun to support organic farming through agricultural subsidy reform. Organic production and marketing have grown at a fast pace.

The term "organic" can be broadly described as food grown without the assistance of man-made chemicals. The beginnings of the organic movement can be traced back to the beginning of the 1800s. In 1840 Justus Von Liebig developed a theory of mineral plant nutrition. Liebig believed that manure could be directly substituted for mineral salts. Many years later in 1910, preceding the First World War, chemists Fritz Haber and Carl Bosh developed an ammonia synthesis process, making use of nitrogen from the atmosphere. This form of ammonia had already been used to manufacture explosives, so after the war, it was implemented into the fertilization of agriculture.

Organic food was initially seen as a fad observed by the eccentric few, however today it has become more widespread. "Organics have come to represent a safe house in a disturbing world where food quality and safety are constantly under siege" (Blythman). Today, whole foods stores have captured a significant share of the grocery shopping market, specifically, Whole Foods Market, Wild Oats, and others.

Structure

Specifications for what may be classified as "organic" food may vary by location. For all intents and purposes, the following quote from USDA.gov presents a fair structure for the movement.

"What is organic food? Organic food is produced by farmers who emphasize the use of renewable resources and the conservation of soil and water to enhance environmental quality for future generations. Organic meat, poultry, eggs, and dairy products come from animals that are given no antibiotics or growth hormones. Organic food is produced without using most conventional pesticides; fertilizers made with synthetic ingredients or sewage sludge; bioengineering; or ionizing radiation. Before a product can be labeled 'organic,' a Government-approved certifier inspects the farm where the food is grown to make sure the farmer is following all the rules necessary to meet USDA organic standards. Companies that handle or process organic food before it gets to your local supermarket or restaurant must be certified, too."

People

According to ResearchWikis, the purchasing of organic food stuffs in has risen every year since 1990 to latest record breaking year. Purchasing continues to gain among all income groups and races, and is gaining the most ground with women ages 25-35, white, & married demographic.

One of the pioneers was John Battendieri.

Battendieri pressed his first batch of "Mr. Natural" apple juice in 1972, making it one of the first packaged organic products.

Organic Companies

The recent interest in the organic industry has sparked the interest of many business. Among these businesses are large companies that seek to distribute many products in large amounts to the American public. The organic market is now a 14 billion dollar a year industry, that continues to grow especially from large corporations such as Wal-Mart that are now offering organic choices to their customers. Other companies leading the organic revolution include food companies such as; General Mills and Kraft that are making the switch to offer more organic option for customers. Along with major companies now offering more organic choices many large companies are buy up smaller already established organic companies such as Earth's Best, Rice Dream soy milk, Garden of Eatin', Celestial Seasonings and Health Valley. By having large companies buy up the already established smaller organic companies it exposes these companies to the popular culture, therefore making them popular because they are in large quantities.

Organic Cosmetics

Cosmetic products that are made with organic ingredients are made without the use of harsh chemicals like pesticides, herbicides, fungicides, and fertilizers. The Environment Protection Agency (EPA) approved pesticides to be used long before research was done that now has linked these chemicals to cancer, and other diseases. Many manufacturers are misleading when it comes to labels on the back of organic cosmetics. For example, manufacturers may use the word "organic" its scientific meaning of "containing carbon."

In order for cosmetics to truly be organic, the Organic Consumers Association suggests that all the cleaning and conditioning ingredients be made from organic materials, the manufacturing process should be simple and ecologica. Non-agricultural water like floral water or botanical water should not be used on labelling because the majority of floral waters used in the cosmetics are the water by-product of essential oil distillation after all the essential oil has been removed. Companies unfortunately mislead consumers by listing floral waters and botanical water extracts and infusions first, burying the synthetic surfactant ingredients that in fact make up the core of the product further down on the labelling.

However some floral waters in skin care products have been created using a distillation process using water only (which takes longer and is usually more expensive to buy and create) and no alcohol or synthetic surfactant ingredients are used to create this, therefore proving to be legitimately organic and created using a natural process.

Organic Farming

Organic farming methods have proven to be more effective than many conventional farmers realised, but they are still placed at a relative disadvantage by their lack of synthetic nitrogen. Thus they rely on price premiums, but these price premiums often allow them to reap higher profits than conventional farms.

Criticisms of the Organic Movement

The organic movement has transformed the mentality that the general public held about food. However, not everyone is a fan of organic products, higher prices for instances are one of the first things critics attack. Originally, organic farming was characterised by family farming, which contributed to its wholesome image. Initially, organic products were for a small portion of the population, the portion that is left over after what society determined was popular culture. Over the years the idea of organic food and products caught on, becoming more popular and widespread. The event of organics becoming more normative marked the transition from high culture to popular culture. Today, organic farming is not representative of quality family farming, it is dominated by big corporations and has become big business. "A single company controls 70 percent of the market in organic milk, California grows about 400 million dollars per year in organic produce and the Organic Trade association includes the biggest names in agribusiness, such as Archer Daniels Midland, Gerber, and Heinz " (Miller, 9). Despite the modest beginnings of the organic movement, organic farming is now a lucrative source of revenue, and its humble beginnings are no longer representative of the organic products on the market today.

Some organic food advocates claim that organic food tastes better, though whether or not it actually does is debatable. The organic movement emphasizes eating local produce that is in season, which in turn, produces fresher food. Essentially, the food being tastier is not attributable to the fact that it's organic (Miller, 9). "Organic food may be fresh, but it's fresh from the manure fields" (Miller, 10).

Critics further protest against the idea that organic food is healthier. Miller, the author of The Organic Myth asserts, "organic food is no more nutritious than food fuelled by industrial fertilizers that are sprayed with synthetic pesticides and genetically altered in science labs." He goes on to say that bacteria are a source of disease in organic food and organic farming is incredibly inefficient (Miller).

Supporters of the Organic Movement

Proponents of organic agriculture point to the fact that organically-grown foods contain higher nutrient levels than those grown with synthetic pesticides and herbicides, and the environmental benefits of not using these often persistent and toxic chemicals. Organic farming also promotes greater biodiversity on farms, as the lack of poison allows species other than those being cultivated to inhabit the farms.

Groups such as IFOAM are active supporters of the organic movement throughout the world. They readily state their goals and are taking their idea to the world.

"Agriculture is one of humankind's most basic activities because all people need to nourish themselves daily. History, culture and community values are embedded in agriculture. The Principles apply to agriculture in the broadest sense, including the way people tend soils, water, plants and animals in order to produce, prepare and distribute food and other goods. They concern the way people interact with living landscapes, relate to one another and shape the legacy of future generations."

This organisation and others like it come under attack from political right in the US and other countries, but it has been accredited by many United Nations organisations.

"Several of the management practices evolved by the organic farming movement have a direct relevance to the battle to achieve food security. Examples include diversification and crop rotation, as well as the use of natural means to combat pests."

Quote from Jacque Diouf, Director General of the FAO in IFOAM's 25th anniversary issue of its magazine Ecology & Farming.

IFOAM has been an FAO-accredited international organisation since 1997.

In March 2002, IFOAM obtained observer status with UNCTAD, and in addition received classification in a special category to participate in the meetings of the Commission on Trade on Goods and Services and Commodities, the Commission on Enterprise, Business Facilitation and Development and the meetings of the UNCTAD Board.

"Organic agriculture can play an important role to achieve the goals of the Convention on Biodiversity, in the global context of the

Millennium Development Goals, by significantly reducing the current rate of biodiversity loss. In addition, organic farming offers direct economic benefits to resource-poor farmers and thereby makes an important contribution to both poverty alleviation and sustainable development."

-Statement made by UNEP Executive Director Klaus Toepfer at the IFOAM Conference on Biodiversity and Organic Agriculture that UNEP co-hosted September 24–26, 2004 in Nairobi, Kenya.

IFOAM is an accredited international organisation with UNEP, but that only one aspect of a more complex relationship between the two organisations.

In September 2004, UNEP hosted and jointly organised the Third International IFOAM Conference on Biodiversity and Organic Agriculture at the headquarters of UNEP in Nairobi, Kenya.

IFOAM and the UNEP are in the process of developing a three-year joint work program for biodiversity and organic agriculture in order to implement the recommendations resulting from the conference.

With organisations such as these gaining in their presence worldwide organic farming practices will likely gain ground in the foreseeable future.

Timeline

- In the summer of 1924 Rudolf Steiner presented what has been called the first organic agriculture course to a group of over one hundred farmers and others at Koberwitz, now Kobierzyce, Poland. In Germany Rudolf Steiner's Spiritual Foundations for the Renewal of Agriculture, published in 1924, led to the popularisation of biodynamic agriculture, probably the first comprehensive organic farming system, that was based on Steiner's spiritual and philosophical teachings.
- The first use of the term "organic farming" is by Lord Northbourne (aka Walter James, 4th Baron Northbourne). The term derives from his concept of "the farm as organism", which he expounded in his book, Look to the Land (1940), and in which he described a holistic, ecologically balanced approach to farming.
- Sir Albert Howard's 1940 book, *An Agricultural Testament*, was influential in promoting organic techniques, and his 1947 book "The Soil and Health, A Study of Organic Agriculture"

adopted Northbourne's terminology and was the first book to include "organic" agriculture or farming in its title.

- In 1939, strongly influenced by Sir Howard's work, Lady Eve Balfour launched the Haughley Experiment on farmland in England. It was the first, side-by-side comparison of organic and conventional farming. Four years later, she published *The Living Soil*, based on the initial findings of the Haughley Experiment. It was widely read, and lead to the formation of a key international organic advocacy group, the Soil Association.
- During the 1950s, sustainable agriculture was a research topic of interest. The science tended to concentrate on the new chemical approaches. In the U.S., J.I. Rodale began to popularise the term and methods of organic growing. In addition to agricultural research, Rodale's publications through the Rodale Press helped to promote organic gardening to the general public.
- In 1962, Rachel Carson, a prominent scientist and naturalist, published *Silent Spring*, chronicling the effects of DDT and other pesticides on the environment. A bestseller in many countries, including the US, and widely read around the world, *Silent Spring* was instrumental in the US government's 1972 banning of DDT. The book and its author are often credited with launching the environmental movement.
- In the 1970s, worldwide movements concerned with environmental pollution caused by persistent agrichemical increased attention on organic farming. One goal of the *organic movement* was to promote consumption of locally grown food, which was promoted through slogans such as *"Know Your Farmer, Know Your Food"*.
- In 1972, the International Federation of Organic Agriculture Movements (IFOAM), was founded in Versailles, France. IFOAM was dedicated to the diffusion of information on the principles and practices of organic agriculture across national and linguistic boundaries.
- In the 1980s, around the world, various farming and consumer groups began seriously pressuring for government regulation of organic production to ensure standards of production. This led to various legislation and certification standards being enacted through the 1990s and to date. Currently, most aspects

of organic food production are government-regulated in the US and the European Union.

- In the 2000s, the worldwide market for organic products (including food, beauty, health, bodycare, and household products, and fabrics) has grown rapidly. Monitoring and challenging certification rules and decisions have become a regular, high profile aspect of activists in the organic movement.

Seasonal Food

Seasonality of food refers to the times of year when a given type food is at its peak, either in terms of harvest or its flavour. This is usually the time when the item is the cheapest and the freshest on the market. The food's peak time in terms of harvest usually coincides with when its flavour is at its best. There are some exceptions; an example being sweet potatoes which are best eaten quite a while after harvest.

History

Seasonal food was practiced since ancient civilizations as people ate what nature produced which varied according to seasons.

In 8th century, however, the choice of what to eat in every season became a conscious social event. Cordoba Calendar, a historical record written in mid 10th century, provided unique detailed information about the eating habits of Spain under the Muslim rule. Typical winter meals were based on rich vegetables such as Seakale beet, cauliflower, turnips, parsnips, carrots, celery, coriander, peas, broad beans, lentils, chickpeas, olives, hard wheat (burghol), couscous, pasta, walnut, almonds, pistachio, and pine kerneis. These were usually taken with meat based diet included lamb, camel and trotters, while fruity desserts consisted of dried figs, dates, raisins, and prunes, accompanied with drinks from syrups violet, jasmine, aloes, medicament spices, fruit pastilles and gums.

In contrast, summer diet consisted of green beans, radish, lettuces, chicories, aubergine, carrots, cucumber, gherkins, watercress, marrow, courgettes, and rice. The meat accompanied these vegetables consisted mainly of poultry, ostrich and beef products. Fruity deserts included fruits such as lemon, lime quinces, nectarines, mulberry, cherries, plums, apricot, grapes, pomegranates, watermelon, pears, apple, and melon. Meanwhile, the drinks involved syrups and jams. Fruit pastels, lemon, rose, jasmine, ginger and fennel.

In Autumn, meals included cabbage, cauliflower, carrots, celery, gourd, wheat, barley, millet, turnips, parsnips, onions, acorns, pulses, and olive oil. Drinks incorporated aromatic herbs and flower distillations of essential oils.

In Spring, meals consisted of onions, gourd, spinach roquette salad, asparagus, lettuces, marrow, fennel, artichokes, fresh broad beans, lemons, cardoons, truffles, peas, wild artichokes, beetroot, basil, mint, sweet marjoram, saffron, green barley, pigeons, lamb and dairy products. Drinks involved lemon and mint syrup, distillation of orange blossom, rose and other herbs for winter.

There has been considerable controversy about how far people should be encouraged to eat seasonal food. In 2008, the chef Gordon Ramsay attracted media coverage when he stated that restaurants should be fined for serving non-seasonal food. In September and October 2008, Valentine Warner presented a programme for BBC Two, entitled *What to Eat Now*, persuading people to consume seasonal food.

Spring

Vegetables:

- Artichoke
- Asparagus
- Bamboo shoot
- Carrots
- Cauliflower
- Celery
- Kale
- New potato
- Onion
- Parsley
- Radish
- Rhubarb
- Samphire or sea grass
- Spinach
- Watercress.

Fruit:

- Raspberry
- Strawberry.

Seafood:

- Clam
- Cockles
- Flying fish
- Kokanee
- Red sea bream
- Salangidae
- Sardines
- Skipjack tuna.

Game:

- Grouse
- Hare
- Chicken.

Summer

Vegetables:

- Eggplant
- Green Bean
- Baby Turnip
- Zucchini/Courgette
- Bell Pepper
- Samphire
- Basil
- Cannellini bean
- Trompettes de Mort
- Broad Bean
- Runner bean
- Swiss Chard
- Peas
- New potato
- Borlotti bean
- Sweetcorn cobs
- Chanterelle (mushrooms)
- Cucumber

- Okra
- Bitter melon.

Fruit:

- Cherry
- Greenguages
- Blackberries
- Nectarines
- Apricot
- Mango
- Victoria plums
- Watermelon
- Fig
- Strawberry
- Peach
- Rhubarb
- Blackcurrant
- Raspberry
- Walnut
- Cantaloupe Melon
- Ume
- Tomato.

Seafood:

- Soft shell crab
- Crayfish
- Sea Bass
- Sea Trout
- Sardines
- Clams
- Skipjack tuna
- Aji
- Rainbow trout
- Ayu
- Unagi (Japanese eel)
- Pacific jumping squid.

Autumn (Fall)

Vegetable: Carrot, Chestnut, Pumpkin, Sweet potato, Lettuce, Renkon, Turnip, Shiitake mushrooms (though they are not technically vegetables.)

Fruit:

- Apple, Pear, Grape, Fig.

Seafood:

- Mackerel, Salmon, Oyster, Pacific saury.

Game:

- Big game, e.g. deer, elk, moose, etc.

Meat:

- Lamb.

Winter

Vegetables:

- Carrot, Daikon, Chinese cabbage, Swede (or Rutabaga), Celeriac, Turnip, Brussel Sprout, Pumpkin, Beetroot, Parsnip, Red Cabbage, Leek, Cabbage, Shallot

Fruits:

- Pear, Tangerine.

Game:

- Duck, Goose

Seafood:

- Sea bass (Grouper), Scallop, Pacific yellowtail, Pacific cod, Monkfish, Halibut, Lobster, Mussels, Stone Crab

Sustainable Agriculture

Sustainable agriculture is the practice of farming using principles of ecology, the study of relationships between organisms and their environment. It has been defined as "an integrated system of plant and animal production practices having a site-specific application that will last over the long term:

- Satisfy human food and fibre needs
- Enhance environmental quality and the natural resource base upon which the agricultural economy depends
- Make the most efficient use of non-renewable resources and on-farm resources and integrate, where appropriate, natural

biological cycles and controls

- Sustain the economic viability of farm operations
- Enhance the quality of life for farmers and society as a whole."

Sustainable agriculture in the United States was addressed by the 1990 farm bill. More recently, as consumer and retail demand for sustainable products has risen, organisations such as Food Alliance and Protected Harvest have started to provide measurement standards and certification programs for what constitutes a sustainably grown crop.

Farming and Natural Resources

The physical aspects of sustainability are partly understood. Practices that can cause long-term damage to soil include excessive tillage (leading to erosion) and irrigation without adequate drainage (leading to salinization). Long-term experiments have provided some of the best data on how various practices affect soil properties essential to sustainability. There is a federal agency, USDA-Natural Resources Conservation Service that specialises in providing technical and financial assistance for those interested in pursuing natural resource conservation and production agriculture as compatible goals.

The most important factors for an individual site are sun, air, soil and water. Of the four, water and soil quality and quantity are most amenable to human intervention through time and labour.

Although air and sunlight are available everywhere on Earth, crops also depend on soil nutrients and the availability of water. When farmers grow and harvest crops, they remove some of these nutrients from the soil. Without replenishment, land suffers from nutrient depletion and becomes either unusable or suffers from reduced yields. Sustainable agriculture depends on replenishing the soil while minimising the use of non-renewable resources, such as natural gas (used in converting atmospheric nitrogen into synthetic fertilizer), or mineral ores (e.g., phosphate). Possible sources of nitrogen that would, in principle, be available indefinitely, include:

1. recycling crop waste and livestock or treated human manure
2. growing legume crops and forages such as peanuts or alfalfa that form symbioses with nitrogen-fixing bacteria called rhizobia
3. industrial production of nitrogen by the Haber Process uses hydrogen, which is currently derived from natural gas, (but this hydrogen could instead be made by electrolysis of water using electricity (perhaps from solar cells or windmills)) or

4. genetically engineering (non-legume) crops to form nitrogen-fixing symbioses or fix nitrogen without microbial symbionts.

The last option was proposed in the 1970s, but is only recently becoming feasible. Sustainable options for replacing other nutrient inputs (phosphorus, potassium, etc.) are more limited.

More realistic, and often overlooked, options include long-term crop rotations, returning to natural cycles that annually flood cultivated lands (returning lost nutrients indefinitely) such as the Flooding of the Nile, the long-term use of biochar, and use of crop and livestock landraces that are adapted to less than ideal conditions such as pests, drought, or lack of nutrients.

Crops that require high levels of soil nutrients can be cultivated in a more sustainable manner if certain fertilizer management practices are adhered to.

Water

In some areas, sufficient rainfall is available for crop growth, but many other areas require irrigation. For irrigation systems to be sustainable they require proper management (to avoid salinization) and must not use more water from their source than is naturally replenished, otherwise the water source becomes, in effect, a non-renewable resource. Improvements in water well drilling technology and submersible pumps combined with the development of drip irrigation and low pressure pivots have made it possible to regularly achieve high crop yields where reliance on rainfall alone previously made this level of success unpredictable. However, this progress has come at a price, in that in many areas where this has occurred, such as the Ogallala Aquifer, the water is being used at a greater rate than its rate of recharge.

Several steps should be taken to develop drought-resistant farming systems even in "normal" years, including both policy and management actions: 1) improving water conservation and storage measures, 2) providing incentives for selection of drought-tolerant crop species, 3) using reduced-volume irrigation systems, 4) managing crops to reduce water loss, or 5) not planting at all.

Indicators for sustainable water resource development are: £ Internal renewable water resources. This is the average annual flow of rivers and groundwater generated from endogenous precipitation, after ensuring that there is no double counting. It represents the maximum amount of water resource produced within the boundaries of a country. This value, which is expressed as an average on a yearly basis, is invariant in time (except in the case of proved climate change). The indicator can be expressed in three different units: in absolute terms (km3/yr), in mm/yr (it is a measure of the humidity of the country), and as a function of population (m3/person per yr).

- Global renewable water resources. This is the sum of internal renewable water resources and incoming flow originating outside the country. Unlike internal resources, this value can vary with time if upstream development reduces water availability at the border. Treaties ensuring a specific flow to be reserved from upstream to downstream countries may be taken into account in the computation of global water resources in both countries.
- Dependency ratio. This is the proportion of the global renewable water resources originating outside the country, expressed in percentage. It is an expression of the level to which the water resources of a country depend on neighbouring countries.
- Water withdrawal. In view of the limitations described above, only gross water withdrawal can be computed systematically on a country basis as a measure of water use. Absolute or per-person value of yearly water withdrawal gives a measure of the importance of water in the country's economy. When expressed in percentage of water resources, it shows the degree of pressure on water resources. A rough estimate shows that if water withdrawal exceeds a quarter of global renewable water resources of a country, water can be considered a limiting factor to development and, reciprocally, the pressure on water resources can have a direct impact on all sectors, from agriculture to environment and fisheries.

Soil

Soil erosion is fast becoming one of the worlds greatest problems. It is estimated that "more than a thousand million tonnes of southern Africa's soil are eroded every year. Experts predict that crop yields will be halved within thirty to fifty years if erosion continues at present rates." Soil erosion is not unique to Africa but is occurring worldwide. The phenomenon is being called *Peak Soil* as present large scale factory farming techniques are jeopardizing humanity's ability to grow food in the present and in the future. Without efforts to improve soil management practices, the availability of arable soil will become increasingly problematic.

Some Soil Management techniques

1. No-till farming
2. Keyline design
3. Growing wind breaks to hold the soil
4. Incorporating organic matter back into fields
5. Stop using chemical fertilizers (which contain salt)
6. Protecting soil from water runoff.

Economics

Socioeconomic aspects of sustainability are also partly understood. Regarding less concentrated farming, the best known analysis is Netting's study on smallholder systems through history. The Oxford Sustainable Group defines sustainability in this context in a much broader form, considering effect on all stakeholders in a 360 degree approach

Given the finite supply of natural resources at any specific cost and location, agriculture that is inefficient or damaging to needed resources may eventually exhaust the available resources or the ability to afford and acquire them. It may also generate negative externality, such as pollution as well as financial and production costs.

The way that crops are sold must be accounted for in the sustainability equation. Food sold locally does not require additional energy for transportation (including consumers). Food sold at a remote location, whether at a farmers' market or the supermarket, incurs a different set of energy cost for materials, labour, and transport.

Methods

What grows where and how it is grown are a matter of choice. Two of the many possible practices of sustainable agriculture are crop rotation and soil amendment, both designed to ensure that crops being cultivated can obtain the necessary nutrients for healthy growth. Soil amendments would include using locally available compost from community recycling centres. These community recycling centres help produce the compost needed by the local organic farms.

Many scientists, farmers, and businesses have debated how to make agriculture sustainable. Using community recycling from yard and kitchen waste utilises a local area's commonly available resources.

These resources in the past were thrown away into large waste disposal sites, are now used to produce low cost organic compost for organic farming. Other practices includes growing a diverse number of perennial crops in a single field, each of which would grow in separate season so as not to compete with each other for natural resources. This system would result in increased resistance to diseases and decreased effects of erosion and loss of nutrients in soil. Nitrogen fixation from legumes, for example, used in conjunction with plants that rely on nitrate from soil for growth, helps to allow the land to be reused annually. Legumes will grow for a season and replenish the soil with ammonium and nitrate, and the next season other plants can be seeded and grown in the field in preparation for harvest.

Monoculture, a method of growing only one crop at a time in a given field, is a very widespread practice, but there are questions about its sustainability, especially if the same crop is grown every year. Today it is realised to get around this problem local cities and farms can work together to produce the needed compost for the farmers around them. This combined with growing a mixture of crops (polyculture) sometimes reduces disease or pest problems but polyculture has rarely, if ever, been compared to the more widespread practice of growing different crops in successive years (crop rotation) with the same overall crop diversity. Cropping systems that include a variety of crops (polyculture and/or rotation) may also replenish nitrogen (if legumes are included) and may also use resources such as sunlight, water, or nutrients more efficiently (Field Crops Res. 34:239).

Replacing a natural ecosystem with a few specifically chosen plant varieties reduces the genetic diversity found in wildlife and makes the organisms susceptible to widespread disease. The Great Irish Famine (1845–1849) is a well-known example of the dangers of monoculture. In practice, there is no single approach to sustainable agriculture, as the precise goals and methods must be adapted to each individual case. There may be some techniques of farming that are inherently in conflict with the concept of sustainability, but there is widespread misunderstanding on impacts of some practices. Today the growth of local farmers' markets offer small farms the ability to sell the products that they have grown back to the cities that they got the recycled compost from. By using local recycling this will help move people away from the slash-and-burn techniques that are the characteristic feature of shifting cultivators are often cited as inherently destructive, yet slash-and-burn cultivation has been practiced in the

Amazon for at least 6000 years; serious deforestation did not begin until the 1970s, largely as the result of Brazilian government programs and policies. To note that it may not have been slash-and-burn so much as slash-and-char, which with the addition of organic matter produces terra preta, one of the richest soils on Earth and the only one that regenerates itself.

There are also many ways to practice sustainable animal husbandry. Some of the key tools to grazing management include fencing off the grazing area into smaller areas called paddocks, lowering stock density, and moving the stock between paddocks frequently.

Several attempts have been made to produce an artificial meat, using isolated tissues to produce it in vitro; Jason Matheny's work on this topic, which in the New Harvest project, is one of the most commented.

Soil Treatment

Soil steaming can be used as an ecological alternative to chemicals for soil sterilization. Different methods are available to induce steam into the soil in order to kill pests and increase soil health. Community and farm composting of kitchen, yard, and farm organic waste can provide most if not all the required needs of local farms. This composting could potentially be a reliable source of energy.

Off-farm Impacts

A farm that is able to "produce perpetually", yet has negative effects on environmental quality elsewhere is not sustainable agriculture. An example of a case in which a global view may be warranted is over-application of synthetic fertilizer or animal manures, which can improve productivity of a farm but can pollute nearby rivers and coastal waters (eutrophication). The other extreme can also be undesirable, as the problem of low crop yields due to exhaustion of nutrients in the soil has been related to rainforest destruction, as in the case of slash and burn farming for livestock feed.

Sustainability affects overall production, which must increase to meet the increasing food and fibre requirements as the world's human population expands to a projected 9.3 billion people by 2050. Increased production may come from creating new farmland, which may ameliorate carbon dioxide emissions if done through reclamation of desert as in Palestine, or may worsen emissions if done through slash and burn farming, as in Brazil. Additionally, Genetically modified organism crops show promise for radically increasing crop yields, although many people and governments are apprehensive of this new farming method.

Urban Planning

There has been considerable debate about which form of human residential habitat may be a better social form for sustainable agriculture.

Many environmentalists advocate urban developments with high population density as a way of preserving agricultural land and maximising energy efficiency. However, others have theorised that sustainable ecocities, or ecovillages which combine habitation and farming with close proximity between producers and consumers, may provide greater sustainability.

The use of available city space (e.g., rooftop gardens, community gardens, garden sharing, and other forms of urban agriculture) for cooperative food production is another way to achieve greater sustainability.

One of the latest ideas in achieving sustainable agricultural involves shifting the production of food plants from major factory farming operations to large, urban, technical facilities called vertical farms. The advantages of vertical farming include year-round production, isolation from pests and diseases, controllable resource recycling, and on-site production that reduces transportation costs. While a vertical farm has yet to become a reality, the idea is gaining momentum among those who believe that current sustainable farming methods will be insufficient to provide for a growing global population.

Wild Culture

Wild culture is the umbrella term used to include all aspects and styles of "hunting and gathering" food harvesting. Wildculture - harvesting the bounty of nature - was the method of food gathering prior to the development of agriculture thousands of years ago. Some wildculture has persisted side by side with agriculture, with some cultures and communities using both styles of food acquisition, and a few reliant solely on wildculture.

As well as organic standards for agriculture, there are now standards for organic wildculture. The total hectares of wild harvested organic land now exceeds the total of certified organic agriculture land. Finland, Zambia and Brazil lead the world, and together account for more than half of the 33.8 million hectares of certified organic wildculture land reported in 2008.

Wildculture is not to be confused with the term "wild culture", first used in 1986 by the Canadian artist, Whitney Smith (multidisciplinary artist), to refer to "the articulated ecotone between what humans do and what they can't control in nature". Through his experiences foraging unfurled ostrich ferns and other wild foods in the southern Ontario forest, he was inspired to explore this concept through the medium of a literary magazine, The Journal of Wild Culture. It was published in Toronto by the Society for the Preservation of Wild Culture from 1986-1991, and followed by the book, "Wild Culture: Ecology and Imagination", edited by Whitney Smith and Christopher Lowry. There is much press coverage from the period recording Smith and other collaborators on the subject. The journal published two editions of a "Wild Foods Field Guide" in 1988 and 1990, plus wild food recipes in various issues.

Australian author Juleigh Robins has advocated the use and consumption of wild food and presents a series of recipes for introducing the flavours, looks, and textures of Australian wild foods.

Chapter 5

Organic Farming and Gardening: Systems and Approaches

Organic horticulture is the science and art of growing fruits, vegetables, flowers, or ornamental plants by following the essential principles of organic agriculture in soil building and conservation, pest management, and heirloom variety preservation.

The Latin words *hortus* (garden plant) and *cultura* (culture) together form *horticulture*, classically defined as the culture or growing of garden plants. *Horticulture* is also sometimes defined simply as "agriculture minus the plough." Instead of the plough, horticulture makes use of human labour and gardener's hand tools, although some small machine tools like rotary tillers are commonly employed now.

Mulches, cover crops, compost, manures, vermicompost, and mineral supplements are soil-building mainstays that distinguish this type of farming from its commercial counterpart. Through attention to good healthy soil condition, it is expected that insect, fungal, or other problems that sometimes plague plants can be minimised. However, pheromone traps, insecticidal soap sprays, and other pest-control methods available to organic farmers organic gardens are also utilised by organic horticulturists.

Horticulture involves five areas of study. These areas are floriculture (includes production and marketing of floral crops), landscape horticulture (includes production, marketing and maintenance of landscape plants), olericulture (includes production and marketing of vegetables), pomology (includes production and marketing of fruits), and postharvest physiology (involves maintaining quality and preventing spoilage of horticultural crops). All of these can be, and sometimes are, pursued according to the principles of organic cultivation.

Organic horticulture (or organic gardening) is based on knowledge and techniques gathered over thousands of years. In general terms, organic horticulture involves natural processes, often taking place over extended periods of time, and a sustainable, holistic approach - while chemical-based horticulture focuses on immediate, isolated effects and reductionist strategies.

Organic Gardening Systems

There are a number of formal organic gardening and farming systems that prescribe specific techniques. They tend to be more specific than, and fit within, general organic standards. Biodynamic farming is an approach based on the esoteric teachings of Rudolf Steiner. The Japanese farmer and writer Masanobu Fukuoka invented a no-till system for small-scale grain production that he called Natural Farming. French intensive and biointensive methods and SPIN Farming (Small Plot INtensive) are all small scale gardening techniques. These techniques were brought to the United States by Alan Chadwick in the 1930s. This method has since been promoted by John Jeavons, Director of Ecology Action. A garden is more than just a means of providing food, it is a model of what is possible in a community - everyone could have a garden of some kind (container, growing box, raised bed) and produce healthy, nutritious organic food, a farmers market, a place to pass on gardening experience, and a sharing of bounty, promoting a more sustainable way of living that would encourage their local economy. A simple 4' x 8' (32 square feet) raised bed garden based on the principles of bio-intensive planting and square foot gardening uses fewer nutrients and less water, and could keep a family, or community, supplied with an abundance of healthy, nutritious organic greens, while promoting a more sustainable way of living.

Organic gardening is designed to work with the ecological systems and minimally disturb the Earth's natural balance. Because of this organic farmers have been interested in reduced-tillage methods. Conventional agriculture uses mechanical tillage, which is plowing or sowing, which is harmful to the environment. The impact of tilling in organic farming is much less of an issue. Ploughing speeds up erosion because the soil remains uncovered for a long period of time and if it has a low content of organic matter the structural stability of the soil decreases. Organic farmers use techniques such as mulching, planting cover crops, and intercropping, to maintain a soil cover throughout most of the year. The use of compost, manure mulch and other organic fertilizers yields a higher organic content of soils on organic farms and helps limit soil degradation and erosion.

Other methods can also be used to supplement an existing garden. Methods such as composting, or vermicomposting. These practices are ways of recycling organic matter into some of the best organic fertilizers and soil conditioner. Vermicompost is especially easy. The byproduct is also an excellent source of nutrients for an organic garden.

Pest Control Approaches

Differing approaches to pest control are equally notable. In chemical horticulture, a specific insecticide may be applied to quickly kill off a particular insect pest. Chemical controls can dramatically reduce pest populations in the short term, yet by unavoidably killing (or starving) natural control insects and animals, cause an increase in the pest population in the long term, thereby creating an ever increasing problem. Repeated use of insecticides and herbicides also encourages rapid natural selection of resistant insects, plants and other organisms, necessitating increased use, or requiring new, more powerful controls.

In contrast, organic horticulture tends to tolerate some pest populations while taking the long view. Organic pest control requires a thorough understanding of pest life cycles and interactions, and involves the cumulative effect of many techniques, including:

- Allowing for an acceptable level of pest damage
- Encouraging predatory beneficial insects to flourish and eat pests
- Encouraging beneficial microorganisms
- Careful plant selection, choosing disease-resistant varieties
- Planting companion crops that discourage or divert pests
- Using row covers to protect crop plants during pest migration periods
- Rotating crops to different locations from year to year to interrupt pest reproduction cycles
- Using insect traps to monitor and control insect populations.

Each of these techniques also provides other benefits, such as soil protection and improvement, fertilization, pollination, water conservation and season extension. These benefits are both complementary and cumulative in overall effect on site health. Organic pest control and biological pest control can be used as part of integrated pest management (IPM). However, IPM can include the use of chemical pesticides that are not part of organic or biological techniques.

Regenerative Organic Farming: A Solution to Global Warming

Agricultural carbon sequestration has the potential to substantially mitigate global warming impacts. When using biologically based regenerative practices, this dramatic benefit can be accomplished with no de-crease in yields or farmer profits. Even though climate and soil type affect sequestration capacities, these multiple research efforts verify that practical organic agriculture, if practiced on the planet's 3.5 billion tillable acres, could sequester nearly 40 percent of current CO2 emissions.

The Problem with Modern Agriculture

Current farming practices are not sustainable for a number of reasons. Some Midwestern soils that in the 1950s were composed of up to 20 percent carbon are now between 1- and 2-percent carbon. This carbon loss contributes to: soil erosion, by degrading soil structure; increasing vulnerability to drought, by greatly reducing the level of water-holding carbon in the soil; and the loss of soil's native nutrient value.

In addition, prevailing farming practices break down soil carbon into carbon dioxide that is released into the atmosphere, greatly contributing to global warming. Surprising analysis of the nation's oldest continuous cropping test plots in Illinois showed that, contrary to long-held beliefs, nitrogen fertilization does not build up soil organic matter. New data from U.S. government research show that with agriculture using chemical fertilizers and herbicides, the U.S. food system contributes nearly 20 percent of the nation's carbon dioxide emissions. On a global scale, figures from the Intergovernmental Panel on Climate Change (IPCC) say that agricultural land use contributes 12 per cent of global greenhouse gas emissions.

Other negative effects of the modern-farming paradigm include: nutrient overload in our waterways from the use of synthetic nitrogen, loss of energy reserves due to the abundant use of petrol-based chemicals (which put an increasing financial burden on farmers as oil prices rise), degradation of our soils (due to mono-cropping that requires use of synthetic fertilizer for fertility) and animal health and welfare concerns.

Carbon dioxide levels are minimised in summer when lush vegetation promotes a sponging action, and are maximised in winter when plants go dormant. However, the greenhouse gas sponging ability of the soil itself may make more of a difference than what's growing on the land. On a global scale, soils hold more than twice as much carbon (an estimated 1.74 trillion U.S. tons) as does terrestrial vegetation (672 billion U.S. tons). Data from Rodale Institute and other studies indicate that regenerative and organic practices can dramatically alter the carbon storage of arable lands, building soil "humic" substances (also known as soil organic matter) that remain as stable carbon compounds for many years.

The key to greater, more stable carbon sequestration lies in the handling of soil organic matter (SOM). Be-cause SOM is primarily carbon, increases in these levels will be directly correlated with carbon sequestration. While prevailing farming practices using synthetic inputs typically deplete SOM, regenerative farming practices, including the integration of crop and animal production, build it.

Before forests and grasslands were converted to field agriculture, SOM generally composed 6 to 10 percent of the soil volume, well over the 1- to 3-percent levels typical of today's agricultural field systems. Building soil organic matter by better nurturing our agricultural lands can capture the excess atmospheric carbon dioxide and begin returning this lost carbon to the soil. Forests and rangelands hold a greater capacity for carbon sequestration than the above ground biomass measurements often used in equating their values.

Organically managed soils can convert carbon from a greenhouse gas into a food-producing asset. Soils that are rich in carbon conserve water and support healthier plants that are more resistant to drought stress, pests and diseases. Our studies of organic systems have shown an increase of almost 30 percent in- soil carbon over 27 years. The petroleum-based system showed no significant increase in soil carbon in the same time period and some studies have shown that these systems, in fact, may lose carbon.

Researchers are fleshing out the mechanisms by which this soil carbon sequestration takes place. One of the most significant findings is the high correlation between increased soil carbon levels and very high amounts of mycorrhizal fungi. These fungi help slow down the decay of organic matter. Beginning with our Farming Systems Trial, collaborative studies by the USDA's Agriculture Research Service (ARS) led by David Douds, Ph.D., show that the biological support system of mycorrhizal fungi are more prevalent and diverse in organically managed systems than in soils that depend on synthetic fertilizers and pesticides.

These fungi work to conserve organic matter by aggregating organic matter with clay and minerals. In soil aggregates, carbon is more resistant to degradation than in free form and thus more likely to be con-served. These findings demonstrate that mycorrhizal fungi produce a potent glue-like substance called glomalin that stimulates increased aggregation of soil particles. This results in an increased ability of soil to retain carbon. These findings are based on analysis by ARS researchers at the Northern Great Plains Research Lab in Mandan, North Dakota.

In Rodale Institute's FST, soil carbon levels in-creased more in the manure-based organic system than in the legume-based organic system, presumably because the manure stimulates the soil to sequester carbon in more stable forms. The study also showed that soil carbon depends on more than just total carbon additions to the system, be-cause cropping diversity or carbon-to-nitrogen ratios of inputs may also have an effect. We believe the answer lies in the decay rates of soil organic matter under different management systems. The application of soluble nitrogen fertilizers in the petroleum-based system stimulates more rapid and complete decay of organic matter, sending carbon into the atmosphere instead of retaining it in the soil as the organic systems do.

Reducing Emissions, Maintaining yield, Cutting Chemical Run-off

An energy analysis of the FST shows a 33-percent reduction in fossil-fuel use for organic corn/soybean farming systems that use cover crops or compost instead of chemical fertilizer. This translates to less greenhouse gas emissions as farmers adopt more regenerative production methods. Moreover, Rodale Institute's organic rotational no-till system can reduce the fossil fuel needed to produce each no-till crop in the rotation by up to 75 percent compared to standard-tilled organic crops. Research beginning this year at Rodale Institute will compare organic and petroleum-based no-till and tilled systems for the first time within the ongoing FST regime.

Research findings have shown that the biggest energetic input in a conventional, modern corn and soybean system is nitrogen fertilizer for corn, followed by herbicides for both corn and soybean production. The ability of regenerative organic agriculture to be a significant carbon sink and less dependent on fossil-fuel inputs has long-term implications for global agriculture and its role in air-quality policies and programs.

There are economic benefits beyond the reduced input costs to growers. Our FST showed that in all systems, corn and soy-bean yields from the organic systems matched the yields from conventional systems, except in drought years, when regenerative systems yielded about 30 per cent more corn than the petroleum-based system. This yield advantage in drought years is due to the fact that soils higher in carbon can capture more water and keep it available to crop plants.

Further, economic analysis by James Hanson, Ph.D., of the University of Maryland has shown that organic systems in Rodale Institute's FST are competitive in returns with conventional corn and soybean farming—even without market-based organic premiums. These have been consistent for more than a decade, with certified-organic crop prices ranging from 40 to 150 percent higher than standard crop prices.

Farming for carbon capture is also compatible with other environmental and social goals, such as reducing erosion and minimising impact on native ecosystems.

This approach utilises the natural carbon cycle to re-duce the use of purchased synthetic inputs. Because chemical fertilizers and pesticides are not used, nutrient and chemical pollution in waterways is significantly reduced. Not only does this translate into long-term cleaner waterways, but it will also save in environmental cleanup costs at the state and federal level. The immensity of the societal cost of over-fertilization is illustrated by the watersheds feeding into the East Coast's Chesapeake Bay. Despite millions of dollars spent over the past 25 years to help farmers reduce agricultural nutrient losses to the bay, roughly 300 millions pounds of nitrogen (39 percent from agricultural sources) still reaches the bay annually.

Research and Proofs

Data from nearly three decades of research trials indicate that wide-scale implementation of established, scientifically researched and proven practical farming methods will change agriculture from a global warming contributor to a global warming inhibitor, from a problem to a solution.

In the FST organic plots, carbon was sequestered into the soil at the rate of 875 lbs/ac/year in a crop rota-tion utilising raw manure, and at a rate of about 500 lbs/ac/year in a rotation using legume cover crops.

During the 1990s, results from the Compost Utilisation Trial (CUT) at Rodale Institute—a 10-year study comparing the use of composts, manures and synthetic chemical fertilizer—show that the use of composted manure with crop rotations in organic systems can result in carbon sequestration of up to 2,000 lbs/ac/year. By contrast, fields under standard tillage relying on chemical fertilizers lost almost 300 pounds of carbon per acre per year. Storing—or sequestering—up to 2,000 lbs/ac/year of carbon means that more than 7,000 pounds of carbon dioxide are taken from the air and trapped in that field soil.

In 2006, U.S. carbon dioxide emissions from fossil fuel combustion were estimated at nearly 6.5 billion tons. If 7000 lb/CO2/ac/year sequestration rate was achieved on all 434 million acres of cropland in the United States, nearly 1.6 billion tons of carbon dioxide would be sequestered per year, mitigating close to one quarter of the country's total fossil fuel emissions.

This is the emissions-cutting equivalent of taking one car off the road for every two acres under organic re-generative agricultural management, based on a vehicle average of 15,000 miles per year at 23 mpg (U.S. EPA).

Organic agricultural practices are established and have been successfully commercialised, and we believe that these methods are applicable in all scale operations as shown by farmers across the United States—from family truck farms to commercial operations of many thousands of acres.

Four European countries have changed their emission-reduction targets for the Kyoto Protocol to include contributions from organic agriculture policy based on Rodale Institute research. These are: the United Kingdom, the Netherlands, Germany and Denmark. France has recently invited Paul Hepperly, Ph.D.—research director at Rodale Institute as a contributing scientist in their exploration of how organic agricultural practices can be useful in fighting greenhouse gases.

Challenges to Success

Widespread implementation will dramatically benefit from additional support for research and development. For example, more research is needed on the mechanisms responsible for the deep carbon sequestration we see in organically managed agricultural soils and forests. The role of mycorrhizae and glomalin in soil carbon retention requires further investigation, as do other biological mechanisms that result in greater ability to sequester carbon naturally and improve soil properties. While these methods have been replicated in a variety of soils and climates—from California to Senegal—further research is needed to systematically measure carbon-sequestration results in various soils, climates and crops. To date, Rodale Institute's FST and

ARS researchers at Beltsville have studied rotations using mainly grains, and UC Davis has tested cot-ton and tomatoes. Measurement of carbon in soil is also key. For widespread commercialisation, better tools are needed for more predictive, quicker and more precise infield soil-carbon measurement. Rodale Institute is currently testing mineralisation of soil nitrogen as a way to estimate soil carbon levels. Another opportunity that may show great potential is analysis of satellite views of the earth to determine soil-carbon amounts. This approach requires taking into consideration yearly global carbon flux dynamics that track carbon and carbon dioxide flows between atmospheric and biospheric (terrestrial and oceanic) sources, driven in part by seasonal changes of photosynthetic activity.

Knowledge of carbon sequestration in forestry, range and pasture land needs to be combined and evaluated with Rodale Institute's research to gain a global terrestrial perspective on how much carbon could be sequestered to mitigate global warming.

The economic implications of improved soil health, increased biodiversity, improved human health, water savings, stream and bay cleanup, as well as climate change mitigation also need to be evaluated to help shape public policy and international accords.

While research needs are clear, data from research trials and commercial practice have established that the obstacles to nationwide implementation are neither technical nor economic. Rather, the largest obstacles to success are human factors. Public education, cause marketing, retraining—these are the types of programs needed to change behaviours in both farming practices as well as the way people shop and buy. Consumers may be ahead of the market in this case. Demand for organic, no-pesticide and hormone-free products in the United States has increased 20 percent or more each year for the past 14 years. Yet there has only been a 3-percent increase in acres dedicated to organic practices.

Public Education, Training in Organic Regenerative Farming and Public Policy

The current environmental emergency requires a major paradigm shift in the way we provide incentives for our farmers. Incremental changes over a period of many decades are a prescription for continued global warming and other environmental degradation.

Successful implementation of regenerative organic farming practices on a national basis will depend on two factors: a strong bottom-up demand for change, and a top-down shift in state and national policy to support farmers in this transition.

Rodale Institute's experience in training thousands of farmers from around the world has proven that the shift to regenerative farming practices is both doable and practical. It's the decision to change that's hard. Government farm policy must be transformed in a way that incentivizes farmers and drives behavioural change toward wide-scale adoption of regenerative farming practices. Success requires a sustained, multifaceted national public education campaign, training for farmers in regenerative agricultural methods and legislative action.

From a climate change and global warming perspective alone, it would seem imperative that the 2012 Farm Bill replace the system of commodity payments with a program that rewards farmers for conservation and other carbon-enhancing farm practices. Farmers should be paid on the basis of how much carbon they can put into and keep in their soil, not only how many bushels of grain they can produce. Incentives will encourage resource conservation and other carbon-enhancing means of producing crops for food, feed and fibre. The current, antiquated method of paying for a single year's crop would be eliminated.

Rodale Institute's research demands a Farm Bill paradigm shift that invests in environmentally sound systems and monetizes the ecological cost of fossil-fuel use (directly as fuel and indirectly in the manufacture of synthetic inputs for non-regenerative systems). In 2008 global food demand is testing the capacity of petroleum-dependent, export-focused commodity agriculture. This system has not served developed nations as food prices soar—inflamed by biofuel demand and fuel prices—and greenhouse gas emissions increase. It has especially hurt developing nations already struggling with food security issues. Further, U.S. subsidies allow its exported commodity crops to be sold at artificially low prices in foreign markets, running afoul of World Trade Organisation (WTO) provisions for free trade. The United States has lost every major challenge to these "trade distorting" subsidies before the WTO, but has yet to seriously explore the EU approach of "green payments" that support ecological services apart from yield.

The following chart outlines some comparative differences between the current Farm Bill structure—which rewards high-volume production of commodities such as wheat, soybeans, corn and oilseeds—and the proposed carbon-reward system of incentives.

Principles of Organic Agriculture

Organic agriculture has grown out of the conscious efforts by inspired people to create the best possible relationship between the earth and men. Since its beginning the sphere surrounding organic agriculture has become considerably more complex. A major challenge today is certainly its entry into the policy making arena, its entry into anonymous global market and the transformation of organic products into commodities. During the last two decades, there has also been a significant sensitization of the global community towards environmental preservation and assuring of food quality. Ardent promoters of organic farming consider that it can meet both these demands and become the means for complete development of rural areas. After almost a century of neglect, organic agriculture is now finding place in the mainstream of development and shows great promise commercially, socially and environmentally. While there is continuum of thought from earlier days to the present, the modern organic movement is radically different from its original form. It now has environmental sustainability and productivity at its core, in addition to the founders concerns for healthy soil, healthy food and healthy people.

Concept of Organic Farming

Organic farming is very much native to India. This concept of organic farming is based on following principles:

- Nature is the best role model for farming, since it does not use any inputs nor demand unreasonable quantities of water.
- The entire system is based on intimate understanding of nature's ways of replenishment. The system does not believe in mining of the soil of its nutrients and do not degrade it in any way.
- The soil in this system is considered as a living entity
- The soil's living population of microbes and other organisms are significant contributors to its fertility on a sustained basis and must be protected and nurtured, at all cost.
- The total environment of the soil, from soil structure to soil cover is more important and must be preserved.

Definition, Organic farming is a method of farming system, which primarily aims at cultivating the land and raising crops in such a way, so as to keep the soil alive and in good health.

It is the use of organic wastes (crop, animal and farm wastes, aquatic wastes) and other biological materials, mostly produced insitu along with beneficial microbes (biofertilizers) to release nutrients to crops, which connotes the 'organic' nature of organic farming. It is also termed as organic agriculture. In the Indian context it is also termed as 'Javik Krishi'. As per FAO's definition "Organic agriculture is a unique production management system which promotes and enhances agro-ecosystem health, including biodiversity, biological cycles and soil biological activity, and this is accomplished by using on-farm agronomic, biological and mechanical methods in exclusion of all synthetic off-farm inputs".

Organic Agriculture Perspective under Indian Conditions

Organic farming denotes a holistic system of farming which optimises productivity in a sustainable manner through creation of interdependent agri-ecosystems where annual crop plants (e.g. wheat), perennial trees (e.g. horticulture) and animals (including fishes where relevant) are integrated on a given field or property. It is often confused with certified organic farming or labeled organic produce, which is a commercial quality control and marketing mechanism entailing third party certification or any other form of independent certification for individual commodities and the process of their cultivation. It is important to separate the two.

The first - organic farming (OF) is essentially an agricultural activity employing the knowledge/understanding of naturally occurring processes while the second – a certified organic farming (COF) is essentially the same as OF but there is an addition of checks and balances to maintain soil fertility and control insect-pests and diseases. It is done by involving record keeping on various aspects and is dictated by market forces.

About 74% farmers in India are small and marginal farmers. Organic farming is most relevant to them because they are resource poor to provide costly inputs for enhancing yield. Their only resource viz. land, need to be prevented from degradations. In the organic farming system approach, a piece of land is used optimally and to its fullest potential to produce a range of nutritious and healthy food as well as other required commodities in a manner which can healthily feed a small family, and maintain soil health and productivity by agricultural practices based on principles of nature. Pests (both insects and diseases) are also controlled and managed by the selection of crop mixes and using biological control measures.

Emergence of Organic Farming in India

Organic farming has made a credible advancements during the past decade. It is a combined effect of farmers' efforts, NGOs work, Governmental interventions and market forces that organic farming in India has reached a stage where it can swiftly move to occupy prominent space in Indian agriculture. Certified organic farming, which was 42,000 ha during 2003-04, has now grown almost by 20 fold during the last 4 years. As on March 2009, the total cultivated area under organic certification process has crossed one million ha mark. Besides this, there is another 2.5 million ha under certified wild forest harvest collection area. National Project on Organic Farming (NPOF) and National Horticulture Mission (NHM) scheme of Department of Agriculture and Cooperation has significantly contributed to this growth. For quality assurance the country has internationally acclaimed certification process in place for export, import and domestic markets. The National Programme on Organic Production (NPOP) provides necessary policy support. Presently,16 accredited certification agencies are looking after the requirement of certification process and the products certified by them are accepted in many countries including European Union and USA.

Growing Crops under Organic Management

Philosophy: Organic farming package is an integrated approach, where all aspects of farming systems are interlinked with each other and work for each other. A healthy biologically active soil is the source of crop nutrition, on-farm biodiversity controls pests, crop rotation and multiple cropping maintains the system's health and on-farm resource management with integration of cattle ensure productivity and sustainability. Organic management stresses on optimisation of resource use and productivity, rather than maximisation of productivity and over exploitation of resources at the cost of resources meant for future generations.

Management Principals - A living soil is the basis of organic farming. A live, healthy soil with proper cropping pattern, crop residue management and effective crop rotation can sustain optimum productivity over the years, without any loss in soil fertility. Organic farming envisages a comprehensive management approach to improve soil health, the ecosystem of the region and the quality of produce. It includes all agricultural systems that promote environmentally sound production of food and fibres. These systems take local soil fertility as a key to successful production, by respecting the natural capacity of plants, animals and the landscape; they aim to optimise quality in all aspects of agriculture and environment. A living soil can be maintained by continuous incorporation of crop and weed biomass, use of animal dung, urine-based manures (FYM, NADEP, vermicompost), biofertilizers and bioenhancers, special liquid formulations (like vermiwash, compost tea etc) during a crop's duration.

As a thumb rule, crop residues should be returned to the plot, directly or indirectly. Cattle droppings may be returned to the field as compost. As a strategy, the quantity of biomass removed for human food and fibre, cattle feed or firewood from an organic farm should be replaced with any other bio-waste on the farm. But it is important to account for it for preparing the balance sheet of nutrients for each crop being cultivated on the farm. In phosphorous-deficient and acidic soils, some quantity of mineral grade rock phosphate and lime can also be added either by direct application to the field or through addition to compost. The compost can be further enriched by incorporation of biofertilizers, microbial inoculants, etc. Special composts like biodynamic compost, cowpat pit compost, biodynamic preparations such as BD-500 and BD-501, special formulations like Panchgavya, Dashgavya, Biosol etc are also useful and ensure optimum productivity. Use of EM formulation has also been found effective in soil enrichment and compost making. For high nutrient demanding crops and for intermittent soil enrichment use of oilcakes, poultry manure, concentrated manures (mixture of oil cakes, poultry manure and rock phosphate) can also be an ideal low-cost option of manuring.

Important Steps

While turning towards organic it is essential that the basic requirements of the system and the area are properly understood and long term strategies are addressed first. In most parts of the country poor soil health due to loss of organic matter and soil microbial load is a major problem. Reducing water availability and increasing temperature is further adding to the problems. Too much dependence on market for supply of inputs and energy has made the agriculture a cost intensive high input enterprise with diminishing returns. We need to address all these concerns and develop a system, which is not only productive and low cost but also resource conserving and sustainable for centuries to come. To start with, following parameters need to be addressed:

- Enrichment of soil
- Management of temperature

- Conservation of rain water
- Maximum harvesting of sun energy
- Self reliance in inputs
- Maintenance of natural cycles and life forms
- Integration of animals
- Maximum reliance on renewable energy sources, such as solar power and animal power

How to Achieve

1. Enrichment of soil –Use crop residue as mulch, use organic and biological fertilizers, adopt crop rotation and multiple cropping, avoid excessive tilling and keep soil covered with green cover or biological mulch. Do not use any chemicals.
2. Management of temperature - Keep soil covered, Plant trees and bushes on bunds.
3. Conservation of soil and rain water – Dig percolation tanks, maintain contour bunds, farm ponds in sloppy lands and adopt contour row cultivation, maintain low height plantation on bunds.
4. Harvesting of sun energy – Maintain green stand throughout the year through combination of different crops and plantation schedules.
5. Self reliance in inputs – develop your own seed, on-farm production of compost, vermicompost, vermiwash, liquid manures and botanical extracts.
6. Maintenance of life forms – Develop habitat for sustenance of life forms, never use pesticides, create enough diversity.
7. Integration of animals – Animals are important components of organic management and not only provide animal products but also provide enough dung and urine for use in soil.
8. Use of renewable energy – Use solar energy, bio-gas and bullock driven pumps, generator and other machine.

Developing Organic Farm

As organic management is an integrated approach, manipulation and adoption of one or few steps may not yield significant results. For optimisation of productivity all the essential components need to be developed in a systematic manner. These steps include: (i) Habitat development, (ii) on-farm facilities for input production (iii) cropping sequence and combination planning, (iv) 3-4 year rotation plan and (vi) growing of crops suiting to the region, soil and climate.

Development of Farm Facilities and Habitat

Infrastructure – Reserve 3-5% of farm space for utilities, such as space for cattle, vermicompost bed, compost tank, Vermiwash/ compost tea unit etc. 5-7 trees should be planted only on this space, as all utility infrastructure need shade. Irrigation well, water pumping infrastructure etc can also be in this utility area. Dig some percolation tanks (7x3x3mt or of any other size depending upon the rainfall and run-off pattern) for rain water conservation (1 pit per ha) at appropriate places depending upon slope and water flow. If possible develop a farm pond of preferably 20x10 mt size. Keep few 200 lit tanks (1 per acre) for liquid manure preparation and few containers for botanicals. For a 5 acre farm, develop 1-2 vermicompost beds, 1 NADEP tank, 2 biodynamic compost beds, 2- 3 compost tea/ vermiwash units, 5 liquid manure tanks, five cowpat pits and one underground cattle-urine collection tank. Efforts should also be made to produce sufficient quantities of BD-500 (cow horn manure) and BD-501 (cow horn silica). 10-12 horn products are sufficient for 5 acre farm. Use of biodynamic compost prepared with the use of BD-502-507 has also been found to be very effective. Habitat and biodiversity- Management of an appropriate habitat for sustenance of different life forms is an essential component of organic farming. This can be achieved by ensuring crop diversity and by maintaining a wide variety of trees and bushes as per climatic suitability. These trees and bushes will not only ensure the nutrients from air and deep soil layers to surface layer but also attract the birds and predators, friendly insects and also provide the food and shelter. More specifically, if we classify areas into wet and dry farms, then on the wet farms there should be five to six neem trees, one to two wood apples, one to two star fruit, eight to ten guava or sour soap, three to four drumstick, one to two fig and 10–15 bushes of mulberry, star gooseberry, curry leaf etc, and on the dry farms there must be at least five to six neem, one to two bael fruit, eight to ten ber or custard apple, one to two aonla, one to two drumstick and 10–15 bushes of sasaka, nirgundi (Vitex negundo), Cassia auriculata, C. tora, etc.

In hilly areas, Alnus nepalensis is considered to be a wonder tree as it fixes good amount of nitrogen. It is being promoted in a cropping system mode particularly in northeastern India. Bushes of Prunus, oak (Quercus glauca), Pinus species along the farm boundary and yarrow (Achillea millifolium), buck wheat (Fagopyrum esculentum), lupin (Lupinus sativus), Himalayan stinging nettle (Urtica parviflora), marigold, etc., in between the plots invite a lot of predators and also attract a large number of pests.

Fruit orchards also need to maintain adequate diversity with at least 3-5 types of fruit plants and few non-fruit trees (as listed above). Major and minor plots should be separated by bunds about 1.5m wide and should be planted with Glyricidia, perennial Sesbania (jayanti), Leucaena leucocephala, cassia siamea, etc. The internal hedgerow should consist of perennial pigeon pea, Crotalaria, seasonal Sesbania, etc. Lops from these trees will provide enough quantity of biologically fixed nitrogen. In between Glyricidia/Sesbania rows insert few plants of pesticidal value such as Adathoda vesica, Vitax nigundo, Calotropis, Datura alba, Ipomea (Besharam) etc. Surrounding the farm or garden, there should be hedgerows or a live fence of coppiced or pollarded, multipurpose, deep-rooted trees and shrubs and medicinal herbs such as Adathoda vasica, Vitex negundo, Jatropha curcas, etc. Ecological diversity is an essential component of any successful organic farming system.

Trees on utility space can be allowed to grow fully. Trees and bushes on farm bunds should be placed randomly at sufficient distance and pruned at repeated intervals. Glyricidia plants should be planted at close spacing on all major bunds and all around the farm. They will act not only as biological fence but also provide biologically fixed nitrogen to soil. A 400 mt long Glyricidia strip can provide 22.5 kg N/ha per year from the year 3 and up to 77 Kg N/ha from year 7 under rainfed conditions. This can be 75-100% higher under irrigated conditions. Three to four harvests can be made under irrigated conditions and two harvests under unirrigated conditions. Never allow them to grow above 5.5 ft to avoid shading effect. Lopping is used as green leaf manure. Simply harvest them and incorporate in soil before sowing or use as mulch.

Conversion of Soil to Organic

Banning of chemicals: It is widely known fact that some biological processes of plants involved in acquiring nutrients such as nitrogen e.g. N2 fixation are generally inhibited by adding Nitrogen fertilizer. Soil scientists generally caution against non-judicious fertilizer use and encourage use of organic compost otherwise it may lead to deficiency of micronutrients. Therefore in organic farming systems there is no place for chemicals.

Low Input Alternative: In first year simultaneously sow three different types of legumes in strips, first of 60 days (like moong), second of 90-120 days (Cow pea or soybean) and third of more than 120 days (red gram) in strips. Apply mixture of Compost and vermicompost (2:1) @ 2.5 ton per acre enriched with 4 kg Azotobacter and 4 kg PSB biofertilizers or 4 kg consortia of customized cultures as basal dose at the time of sowing preferably in furrows below the seeds. Seeds of legumes should be treated with crop specific strains of Rhizobium biofertilizer. Mulch the entire surface with a thick layer of biological mulch and drench the biomass with Jivamrut @ 200 lit per acre. Seedlings will emerge from this layer. If soil is poor in phosphorus then apply 300 kg of low grade mineral rock-phosphate along with the compost. Apply second dose of Jivamrut after 25-30 days of sowing with irrigation water or during rains. To add to diversity 100 plants/ acre of marigold or Hibiscus subdarifa may be planted randomly through out the field. Few seedlings of vegetables such as chillies, tomato, brinjal, etc and rhizomes of turmeric, ginger etc can be planted randomly for home consumption.

Harvest the pods/ fruits and use remaining biomass for mulch. Collect the crop biomass at the end of strips in the form of heaps and drench with Jivamrut. Sow short duration leafy vegetables (such as fenugreek or spinach) in the space vacated by the first and second crop and mulch the surface with treated biomass. Harvest leafy vegetable and grains and incorporate remaining biomass in the soil at appropriate time. In next season apply compost-vermicompost mixture @ 2.5 ton/ha and sow cereal crop with legume as inter or companion crop. After harvest use entire legume and remaining part of cereal crop as mulch. If irrigation facilities are there, take summer legume with some vegetable crop. Recycle entire residue as mulch. Use 3-4 application of liquid manure (such as Jivamruta) during each cropping season for soil application. Now the soil is ready for high value horticultural crops.

High Input Alternative: Incorporate 2.5-3.0 ton compost/ vermicompost or 1.5 ton of biodynamic compost, 500 kg crushed oil cakes, 500 kg rock phosphate, 100 kg neem cake, 5 kg Azotobacter and 5kg PSB biofertilizer or 4 kg consortia of customized cultures in soil through broadcasting or by drilling in furrows below the seeds. Sow 3-4 types of different crops in strips. 40% crop stand should be of legumes. Randomly plant 100-150 marigold and vegetable seedlings for increased diversity. After harvest incorporate entire residue in soil or use as mulch after sowing of the next crop. For second crop also use similar quantities of manures. Use liquid manure (Jivamruta) @ 200lit/acre 3-4 times during cropping season along with irrigation water. For increased productivity 2-3 sprays of vermiwash or vermiwash+cow urine or Panchgavya can also be provided. In fruit orchards cultivate 3-4 types of legume mixtures as mixed or intercrop in inter spaces along with adequate quantity of manures (as specified above). After pod/ grain harvest mulch the entire soil surface with the left over biomass and drench the biomass with 2 applications of Jivamruta. After about 12-18 months the soil will be ready for organic cultivation of any crop combination. For next two-three years, along with any crop incorporate legumes as inter or companion crops. Ensure that crop residue always have at least 30% residue from legumes. Also treat crop residue with liquid manure before incorporating into soil or using as mulch.

Multiple Cropping and Crop Rotation

Mix cropping is the outstanding feature of organic farming in which variety of crops are grown simultaneously or at different time on the same land. In every season care should be taken to maintain legume cropping at least 40%. Mix cropping promotes photosynthesis and avoids the competition for nutrients because different plants draw their nutrients from different depth of soil. The legume fixes atmospheric nitrogen and make available for companion or succeeding crops. Deep rooted plants drew nutrient from deeper layer of soil and bring them to the surface of soil through their leaf fall. So the nutrients leached down to lower strata are further brought back to upper layer by these deep rooted plants. Also help in protecting soil from soil erosion. Farmers should select the crops combination according to their needs and season.

In selecting crop combinations, only compatible crops should be planted, for e.g. maize gets along well with beans and cucumber, tomatoes go well with onions and marigold. On the other hand beans and onions do not go well with each other.

Entire farm should have at least 8-10 types of crops at all the times. Each field/ plot should have at least 2-4 types of crops out of which one should be legume. In case if only one crop is taken in one plot then adjacent plots should have different crops. For maintenance of diversity and pest control randomly plant 50- 150/acre vegetable seedlings for home consumption and 100 plants/acre of marigold (Genda) in all crop fields. Even high nutrient demanding crops such as sugarcane can also be grown with suitable combination of various legume and vegetable crops with optimum productivity.

Crop Rotation

Crop rotation is the back bone of organic farming practices. To keep the soil healthy and to allow the natural microbial systems working, crop rotation is must. Crop rotation is the succession of different crops cultivated on same land. Follow 3-4 years rotation plan. All high nutrient demanding crops should precede and follow legume dominated crop combination.

Rotation of pest host and non pest host crops helps in controlling soil borne diseases and pest. It also helps in controlling weeds. It is better for improving productivity and fertility of soil. Crop rotations help in improving soil structure through different types of root system. Legumes should be used frequently in rotation with cereal and vegetable crops. Green manure crops should also find place in planning rotations. High nutrient demanding crops should always be followed by legume crops and returned back to soil. Some important benefits of crop rotations are:

a. Not all plants have same nutritive needs
b. Soil structure is improved through different types of roots
c. Pest build up is avoided and
d. Rotations help against the build up of weeds.

Under Network Project on Organic Farming (NPOF of ICAR) important cropping systems, which were found economically better or at par with conventional system at different experimental stations in the country have been identified, which are as follows:

- Tomato/ Cabbage – cauliflower – pea and maize – garlic at Bajaura, Himachal Pradesh
- Rice – wheat/ potato/ mustard/ lentil at Ranchi, Jharkhand
- Groundnut – rabi Sorghum, soybean – durum wheat, potato – chick pea, chillil+ Cotton and maize – chick pea at Dharwad, Karnataka
- Soybean – durum wheat/ mustard/ chick pea/ isabgol at Bhopal, M.P.
- Rice – durum wheat/ berseem, rice – potato – Okra and rice – garlic, sorghum – berseem, maize – berseem – maize + coepea and sorghum + cluster bean – oats-cowpea at Ludhiana, Punjab
- Maize – cotton, chillies – onion and brinjal – sunflower at Coimbatore

- Sorghum – pea – okra at Modipuram, Uttar Pradesh
- Carrot/ rice (pre kharif) – rice (kharif), potato/rice (pre kharif) – rice (kharif), tomato/ rice (pre kharif) – rice (kharif), French bean/ rice (pre kharif) – rice (kharif) at Umiam, Meghalaya.

Status of Rich and Live Organic Soil

A fertile and live organic soil ideally should have Organic Carbon (C) between 0.8-1.5%. At any point of time it should have adequate quantity of dry, semi decomposed and fully decomposed organic matter for the use of micro-flora and fauna.

Total microbial load (bacteria, fungi and actinomycetes) should be above 1 x 108 /gm of soil. There should be at least 3-5 earth worms/ cubic ft of soil. There should be enough quantity of small life forms and insects such as ants etc.

Seed/ Planting Material Treatment

In organic management, protection measures are used only in the case of problematic situations. Use of disease free seed stock and resistant varieties is the best option. There is no standard formulation or treatment methodology, available as on today, but farmers use different methods. Few of such innovative seed treating formulations are as follows:

- Hot water treatment at 530C for 20-30 min.
- Cow urine or cow urine-termite mound soil paste
- Beejamrut
- Asphoetida 250gm in one lit. of water for 10 kg seed
- Turmeric rhizome powder mixed with cow urine
- Panchgavya extract
- Dashparni extract
- Trichoderma viride (4gm/kg seed) or Pseudomonas fluorescence (10gm/kg seed).
- Biofertilizers (Rhizobium/ Azotobacter +PSB).

Preparation of Beejamruta – Put 5 kg of fresh cow dung in a cloth bag and suspend in a container filled with water to extract the soluble ingredients of dung. Suspend 50 g lime in 1 lit water separately. After 12 – 16 hours squeeze the bag to collect extract and add 5 lit cow urine, 50 gm virgin forest soil, lime water and 20 lit water. Incubate for 8-12 hours. Filter the contents. The filtrate is used for seed treatment.

Manuring and Soil Enrichment

During conversion period, soil fertility can be improved and maintained initially through use of organic inputs like well decomposed organic manure/ vermicompost, green manure and biofertilizers in appropriate quantity. These organic inputs are used for feeding the soil. Well fed healthy soil rich in microflora and microfauna takes care of the crop nutrient requirement. Plant biomass, FYM, Cattle dung manure, enriched compost, biodynamic compost, Cow-pat-pit compost and vermicompost are key sources of on-farm inputs. Among off-farm inputs, important components are non-edible oil cakes, poultry manure, biofertilizers, mineral grade rock phosphate and lime etc.

Lopping from Glyricidia and other plants grown on bunds, on-farm produced compost and vermicompost, animal dung and urine and crop residue should form the major source of nutrient and concentrated manures such as crushed oil cakes, poultry manure, vegetable market waste compost and other novel preparations such as biodynamic formulations etc can be used in appropriate quantity. Use of high quantities of manures should be avoided. Changing crop rotations and multiple crops ensure better utilisation of resources. Depending upon the type of crop and requirement of nutrients for different crops, the quantity of externally produced inputs is determined.

Application of liquid manure (for soil enrichment) is essential to maintain the activity of microorganisms and other life forms in the soil. 3-4 applications of liquid manure is essential for all types of crops. Vermiwash, compost tea, cow urine, Pachgavya and Biosol etc are excellent growth promoters when used as foliar spray. 3-5 sprays after 25-30 days of sowing ensure good productivity. Use of Biodynamic preparations, such as BD-500 and BD-501 as foliar spray has also been found to be effective in growth promotion.

Use of Biofertilizers and Microbial Cultures

Biofertilizers viz: Rhizobium, Azotobacter, Azospirillum, PSB and Pseudomonas etc have been found to be very effective tools of fertility management and biological nutrient mobilisation. Recently customized consortia of such biofertilizer organisms, better adapted to local climatic conditions have also been developed and are available commercially. Efficiency of such microbial formulations is much higher under no-chemical use situations, therefore application of such inputs need to be ensured under all cropping situations. Some important formulations for soil enrichment.

Preparation of Liquid Manures

Many variants of liquid manures are being used by farmers of different states. Few important and widely used formulations are given below: Sanjivak – Mix 100 kg cow dung, 100 lit cow urine and 500 gm jaggary in 300 lit of water in a 500-lit closed drum. Ferment for 10 days. Dilute with 20 times water and sprinkle in one acre either as soil spray or along with irrigation water. Jivamrut – Mix cow dung 10 kg, cow urine 10 lit, Jaggary 2 kg, any pulse grain flour 2 kg and Live forest soil 1 kg in 200 lit water. Ferment for 5 to 7 days. Stir the solution regularly three times a day. Use in one acre with irrigation water. Amritpani - Mix 10 kg cow dung with 500 gm honey and mix thoroughly to form a creamy paste. Add 250 gm of cow desi ghee and mix at high speed. Dilute with 200 lit water. Sprinkle this suspension in one acre over soil or with irrigation water. After 30 days apply second dose in between the row of plants or through irrigation water.

Panchgavya – Mix fresh cow dung 5 kg, cow urine 3 lit, cow milk 2 lit, curd 2 lit, cow butter oil 1 kg and ferment for 7 days with twice stirring per day. Dilute 3 lit of Panchgavya in 100 lit water and spray over soil. 20 lit panchgavya is needed per acre for soil application along with irrigation water.

Enriched Panchgavya (or Dashagavya) – Ingredients - cow dung 5 kg, cow urine 3 lit, cow milk 2 lit, curd 2 lit, cow deshi ghee 1 kg, sugarcane juice 3 lit, tender coconut water 3 lit, banana paste of 12 fruits and toddy or grape juice 2 lit. Mix cow dung and ghee in a container and ferment for 3 days with intermittent stirring. Add rest of the ingredients on the fourth day and ferment for 15 days with stirring twice daily. The formulation will be ready in 18 days. Sugarcane juice can be replaced wit 500 g jaggery in 3 litres water. In case of non-availability of toddy or grape juice 100g yeast powder mixed with 100 g jaggery and 2 lit of warm water can also be used. For foliar spray 3-4 lit panchgavya is diluted with 100 litre water. For soil application 50 lit panchagavya is sufficient for one ha. It can also be used for seed treatment.

Management of Temperature

Temperature in summer season is quite high and need to be managed. It can be achieved by keeping soil covered with biological mulch. Surface mulch has been reported to conserve soil moisture and improve water use efficiency. In the long term experiment at ICRISAT, it has been reported that mulch applied in this manner on the hottest day of summer (April 30) in 2002 the soil temperature at 5 and 10 cm depth in the mulch applied plots was 6.5 to 7.3o C lower than in control plot (Rupela et al 2005). Temperature control can also be achieved by planting different types of trees like neem, amla, tamarind, gular, zizipus bushes, gliricidia on bunds.

Protection to All Life Forms

Practice of maintaining enough biomass and mulching with crop and weed residue will ensure the protection to all life forms in soil. Another important practice of banning the chemical fertilizers and pesticides in farming definitely helps in protecting the life forms in soil. For the survivability of different life forms the field must have dry organic matter as a food for small insects and small animals in soil, semi decomposed organic matter as food for earthworms and fully decomposed organic matter for micro organisms in the soil at all times. These insects, small animals, earthworms and microorganisms are the tireless natural employees of the soil, wherein small animals and insects feed on the larvae of pests and thus controlling the pest, earthworms makes the soil porous thus creating the more aerobic conditions in soil and also decompose the half digested organic residue and release locked nutrients into soil. Soil rich in organic carbon contain ample quantity of beneficial micro flora which plays an important role in recycling of nutrients and nitrogen fixation, phosphate solubilization and photosynthesis activity, cellulolytic activity. Therefore protection to all life forms in soil should be ensured at all time.

Pest Management

As in organic farming management use of synthetic chemicals are prohibited, the pest management is done by: (i) cultural or agronomic (ii) mechanical (iii) biological or by (iv) organically acceptable botanical extract or some chemicals such as copper sulphate and soft soap etc.

Cultural alternative: Use of disease free seed or stock and resistant varieties are best preventive practice in organic pest management. Maintenance of biodiversity, effective crop rotation, multiple cropping, habitat manipulation and use of trap crops are also effective practices which can keep the population of pests below economical threshold limit (ETL).

Mechanical alternative: Removal of affected plants and plant parts, collection & destruction of egg masses and larvae, installation of bird perches, light traps, sticky coloured plates and pheromone traps are most effective mechanical methods of pest control.

Biological alternative: Use of pest predators and pathogens has also proved to be effective method of keeping pest problem below ETL. Inundative release of Trichogramma sp. @ 40,000 to 50,000 eggs per hectare, Chelonus blackburni @15,000 to 20, 000 per hectare, Apanteles sp.@15,000 to 20,000 per ha and Chrysoperla sp.@ 5,000 per ha., after 15 days & others parasites & predators after 30 days of sowing, can also effectively control pest problem in organic farming .

Use of Biopesticides - Trichoderma viride or T. harazianum or Pseudomonas fluorescence formulation @ 4gm/kg seed either alone or in combination, manage most of the seed borne & soil borne diseases. There are other formulations viz. Beauvaria bassiana, Metarizium anisopliae, Numeria rileyi, Verticillium sp, which are available in the market and can manage their specific host pest. Bacillus thurengensis stenebrionis and B.thurengensis sandigo are effective against coleopterans as well as some other insect species. Bt. has been used in the management of diamond back moth on crucifers and vegetables @ 0.5-1.0 kg. formulation per ha.

Viral biopesticides of baculovirus group viz. granulosis viruses (GV) and nuclear polyhedrosis viruses provided a great scope in plant protection field. Spray of nuclear polyhedrosis viruses (NPV) of Helicoverpa armigera (H) or Spodoptera litura (S) @ 250 larval equivalents are very effective tools to manage the Helicoverpa sp. or Spodoptera sp. respectively.

Botanical Pesticides

Many plants are known to have pesticidal properties and the extract of such plants or its refined forms can be used in the management of pests. Among various plants identified for the purpose, neem has been found to be most effective.

Neem (Azadirachta indica): Neem has been found to be effective in the management of approximately 200 insects, pests and nematodes. Neem is very effective against grasshoppers, leaf hoppers, plant hoppers, aphids, jassids, and moth caterpillars. Neem extracts, are also very effective against beetle larvae, butterfly, moth and caterpillars such as Mexican bean beetle, Colorado potato beetle and diamondback moth. Neem is very effective against grasshoppers, leaf minor and leaf hoppers such as variegated grasshoppers, green rice leaf hopper and cotton jassids. Neem is fairly good in managing beetles, aphids and white flies, mealy bug, scale insects, adult bugs, fruit maggots and spider mites.

Some other Pest Control Formulations

Many organic farmers and NGOs have developed large number of innovative formulations which are effectively used for control of various pests. Although none of these formulations have been subjected to scientific validation but their wide acceptance by farmers speak of their usefulness. Farmers can try these formulations, as they can be prepared on their own farm without the need of any purchases. Some of the popular formulations are listed below: Cow urine – Cow urine diluted with water in ratio of 1: 20 and used as foliar spray is not only effective in the management of pathogens & insects, but also acts as effective growth promoter for the crop.

Fermented Curd Water: In some parts of central India fermented curd water (butter milk or Chaach) is also being used for the management of white fly, jassids aphids etc.

Dashparni Extract: Crush neem leaves 5 kg, Vitex negundo leaves 2 kg, Aristolochia leaves 2 kg, papaya (Carica Papaya) 2 kg, Tinospora cordifolia leaves 2 kg, Annona squamosa (Custard apple) leaves 2 kg, Pongamia pinnata (Karanja) leaves 2 kg, Ricinus communis (Castor) leaves 2 kg, Nerium indicum 2 kg, Calotropis procera leaves 2 kg, Green chilly paste 2 kg, Garlic paste 250 gm, Cow dung 3 kg and Cow Urine 5 lit in 200 lit water ferment for one month. Shake regularly three times a day. Extract after crushing and filtering. The extract can be stored up to 6 months and is sufficient for one acre. Neem-Cow urine extract - Crush 5 kg neem leaves in water, add 5lit cow urine and 2 kg cow dung, ferment for 24 hrs with intermittent stirring, filter squeeze the extract and dilute to 100 lit, use as foliar spray over one acre. Useful against sucking pests and mealy bugs.

Mixed Leaves Extract: Crush 3 kg neem leaves in 10 lit cow urine. Crush 2 kg custard apple leaf, 2 kg papaya leaf, 2kg pomegranate leaves, 2 kg guava leaves in water. Mix the two and boil 5 times at some interval till it becomes half. Keep for 24 hrs, then filter squeeze the extract. This can be stored in bottles for 6 months. Dilute 2-2.5 lit of this extract to 100 lit for 1 acre. Useful against sucking pests, pod/fruit borers.

Chilli-garlic Extract: Crush 1 kg Ipomea (besharam) leaves, 500 gm hot chilli, 500 gm garlic and 5 kg neem leaves in 10 lit cow urine. Boil the suspension 5 times till it becomes half. Filter squeeze the extract. Store in glass or plastic bottles. 2-3 lit extract diluted to 100 lit is used for one acre. Useful against leaf roller, stem/fruit/pod borer.

Broad Spectrum Formulation - 1: In a copper container mix 3 kg fresh crushed neem leaves and 1 kg neem seed kernel powder with 10 lit of cow urine. Seal the container and allow the suspension to ferment for 10 days. After 10 days boil the suspension, till the volume is reduced to half. Ground 500 gm green chillies in 1 lit of water and keep overnight. In another container crush 250gm of garlic in water and keep overnight. Next day mix the boiled extract, chilli extract and garlic extract. Mix thoroughly and filter. This is a broad spectrum pesticide and can be used on all crops against wide variety of insects. Use 250 ml of this concentrate in 15 lit of water for spray.

Broad Spectrum Formulation - 2: Suspend 5 kg neem seed kernel powder, 1kg Karanj seed powder, 5 kg chopped leaves of besharam (Ipomea sp.) and 5kg chopped neem leaves in a 20lit drum. Add 10-12 lit of cow urine and fill the drum with water to make 150 lit. Seal the drum and allow it to ferment for 8-10 days. After 8 days mix the contents and distil in a distiller. Distillate will act as a good pesticide and growth promoter. Distillate obtained from 150lit liquid will be sufficient for one acre. Dilute in appropriate proportion and use as foliar spray. Distillate can be kept for few months without any loss in characteristics. Some Other forms of Organic Management

Biodynamic Agriculture

Biodynamic agriculture is a method of farming that aims to treat the farm as a living system which interacts the environment, to build healthy, living soil and to produce food that nourishes and vitalizes and helps to develop man kind. The underlying principle of biodynamics is making life-giving compost out of dead material. The methods are derived from the teachings of Rudolf Stainer and subsequent practitioners.

The important components of biodynamic farming are as follows:

a. Turning in plant materials such as green crops and straw
b. Not using chemical fertilizers and pesticides
c. Avoiding soil compaction by machinery or animals, particularly in wet weather
d. Keeping soil covered by pasture, crops or mulch not destroying the soil structure by poor farming practices such as excessive use of rotary hoe or cultivation in unsuitable weather (too wet or too dry)
e. Fallowing the land by planting deep-rooting permanent pasture species or using green crops

f. Use of preparations BD-500 and BD-501

g. Compost made with preparations BD-502 – BD-507

h. Liquid manure made with preparations BD-502 – BD-507

i. Cowpat pit manure made with preparations BD-502 – BD-507.

These biodynamic preparations named BD-500 to BD-507 are not food for the plants, but they facilitate the effective functioning of etheric forces. They are also not the usual compost starters, but can stimulate compost organisms in various ways. In short they are biologically active dynamic preparations which help in harvesting the potential of astral and etheral powers for the benefit of the soil and various biological cycles in the soil.

So far 9 biodynamic preparations have been developed, named as formulation 500 to 508. Out of these, formulation-500 (cow horn compost) and formulation-501 (horn-silica) are very popular and are being used by large number of organic farmers. Formulations-502 to 507 are compost enrichers and promoters, while formulation 508 is of prophylactic in nature and helps in control of fungal diseases.

Biodynamic Formulation-500 (BD-500)

As per the established norms of biodynamic process while cow-dung is full of astral and etheral powers; the cow-horn shell has the potential to absorb astral powers. In this formulation the inherent potential of these two components is harvested in making a biologically active formulation. a. Method of preparation - Whip fresh cow dung to prepare a thick smooth paste. Fill the empty cow horn shells with this paste. Now place these horns in the pit in upright position with the pointed closed end of horns facing upwards. Fill the pit with good fertile soil and compost mixture (25: 1) till ground level. The soil of the pit is to be kept moist for all the time. If required sprinkle water at repeated intervals. As per the Indian moon calendar "Kwar Navratra" (October-November) is the most ideal period for placing the dung filled horns in pits. The horns are kept buried for approximately six months and are taken out during "Chaitra Navratra" (March-April). Dig out the horns at appropriate time and take out the BD- 500 compost. The compost should be moist and should have a pleasing smell. Store the compost in earthen pots till its use. BD-500 can be used in any crop twice, first dose is to be applied a day before sowing and second dose after 20 days of seedling emergence. For best results it should be applied close to full moon days. BD-500 applied during lowmoon or no-moon days will not be that effective.

Method of Application – Mix 30 gm of BD-500 in about 13 lit of rain or fresh tube-well water. Stir the solution with hand for one hour. Apply this suspension with the help of Knapsac sprayer on soil surface or as foliar spray. The best time of application is close to sunset. BD-500 application encourage the growth of beneficial microorganisms and earthworms, promote rooting process and harvest terrestrial forces for better crop growth and increased biological activity in the soil.

Biodynamic Formulation 501 (BD-501)

In this formulation fine powder of quartz silica is filled in empty cow-horn shells and kept buried in soil for six months during hot summer season. Opposed to BD-500, the silica filled horns are buried during March-April (Chaitra Navratra) and taken out during Oct.-Nov. (Kwar Navratra). BD-501 is used as foliar spray and is known to be promoting photosynthetic activity of the plants, resulting into better growth of the plants and better quality of grains and fruits. 1gm BD-501 is sufficient for one acre. Mix 1gm BD-501 in 13 lit of water and mix by whirling for one hour. Apply this suspension in the field as fine mist spray. BD-501 should be applied in early morning hours when there is mild breeze. BD-501 is to be applied first at 3-4 leaf stage followed by two more application at an interval of 30 days. BD-501 also acts as prophylactic agent and helps in prevention of many fungal diseases such mildews and blights.

Other Biodynamic Preparation

Besides 500 and 501 there are seven other biodynamic preparations having numbers from 502 to 508, but as their method of preparations are difficult, they are not very popular in India and are not being used in large scale. Their methods of preparation in brief are as follows: BD-502 – Moistened yarrow (Achillea millefolium) blossom gathered in spring, are packed into the bladder of deer stag or hart. The bladder is hung into the sun over the summer and buried into good soil over the winter. The contents, dug up in the spring, will aid the compost to regulate potassium and sulphur processes.

BD-503 – Chamomile blossoms (Matricria chamomilla) gathered in the summer are moistened with chamomile tea and stuffed into the small intestine of a freshly butchered cow, made into little links of sausages and buried into good humus soil in the fall. The burial place should be close to melt water flow of snow after the winter. This preparation helps regulate the calcium processes of compost.

BD-504 – Stinging nettle (Urtica dioica) is buried in the soil for one full year, enclosed in a mantle of peat moss. It aids in humification of the compost.

BD-505 – Scrapings of the outer rind of Oak bark (Quercus robur) are placed in the skull cavity of a domestic animal such as sheep or goat and buried in fall in ground that has water percolations through it (such as below leaking drain pipe). The contents are used in the spring. This preparation works on calcium processes and contributes to making plants disease resistant.

BD-506 – Dried flowers of Dandelion (Taraxacum officinale) gathered in spring are moistened and folded into the mesentery (membrane that holds intestines) of a cow. This is buried in soil until the spring. It helps to regulate the silica processes in relation to the potassium processes.

BD-507 – Extract the juice of Valerian (Valeriana officinalis) flowers by squeezing. The juice is diluted in rain water and sprayed on the compost pile.

This preparation regulates the phosphorus processes in the compost. Although, the method of preparation of these compost BD formulations are difficult, but they are required in very small quantities and can be stored in glass containers for long periods. Once prepared can be used over large number of compost piles. A tea spoonful of each (502-507) will suffice for a normal garden compost pile of 3 m3. On a compost pile poke 5 holes of about 30-40cm deep and stuff each with tea spoonful of formulation 502 to 506.

Formulation 507 is stirred in a bucket of water and uniformly spread over the entire compost pile.

BD-508 – Fresh tissue of horse tail plant (Equisetum arvense) is made into a tea by boiling with water for 20 min. Filtered tea can be stored in glass bottles and diluted at the time of use. This formulation is used as prophylactic agent against mildews, blights and other fungal disorders.

Natural Farming

Natural farming emphasizes on efficient use of on-farm biological resources and enrichment of soil with the use of Jivamruta to ensure high soil biological activity. Use of Bijamruta for seed/ planting material treatment and Jivamruta for soil treatment and foliar spray are important components. The use of both these ingredients have been incorporated in the package described above. Jivamruta has been found to be rich in various beneficial microorganisms. As per the studies conducted by Bio Centre Bangalore the Jivamruta contains following microorganisms:

- Azospirillum 2 x 106
- PSM 2 x 106
- Pseudomonas 2 x 102
- Trichoderma 2 x 106
- Yeasts and moulds 2 x 107

200 lits of jivamruta is needed for one application in one acre. It can be applied through irrigation water by flow, by drip or sprinkler or even by drenching of mulches spread over the field or under the tree basin.

Natueco Farming

The Natueco farming system follows the principles of eco-system networking of nature. It is beyond the broader concepts of organic or natural farming in both philosophy and practice. It offers an alternative to the commercial and heavily chemical techniques of modern farming. Instead, the emphasis is on the simple harvest of sunlight through the critical application of scientific examination, experiments, and methods that are rooted in the neighbourhood resources. It depends on developing a thorough understanding of plant physiology, geometry of growth, fertility, and biochemistry. This can be simply achieved through:

Understanding Natueco Farming Science

- Natueco Farming methods go beyond natural farming and organic farming.
- In natural farming, farming is done trusting nature through the empirical wisdom of ages. However, Natueco methods emphasize farming by knowing nature more and more through critical scientific inquiries and experiments. It is an ever growing, novel, unique, participatory tryst between man and nature. Moreover, Natueco Farming in no way related to the present commercial techniques of farming.
- It has a new vision of infinite resource potentials in Nature and sunlight and promises plenty for all through harvesting all available resources by increasing the human activity.

- This depends on critical understanding of greening and recycling of biomass within the neighbourhood to enrich the structure and fertility of soil in a calculated way.
- It promises record assured yields in a mathematic precision by understanding plant's geometry, cycles of growth and canopy (leaf area) management with little or no external inputs and ensuring optimum harvesting of sunlight.
- It visualizes that in the near future, the present money market system will have to give way to a new eco-economic system of Nature, i.e. energy market system.

Natueco Farming Step by Step

Natueco Farming emphasizes 'Neighbourhood Resource Enrichment' by 'Additive Regeneration' rather than through dependence on external, commercial inputs. The three relevant aspects of Natueco Farming are:

1. Soil - Enrichment of soil by recycling of the biomass by establishing a proper energy chain.
2. Roots - Development and maintenance of white feeder root zones for efficient absorption of nutrients.
3. Canopy - Harvesting the sun through proper canopy management for efficient photosynthesis.

Basic Principals of Natueco Farming

Harvesting the Sun: In all biological processes, energy input is required and solar energy is the only available resource. No time and no square foot of sun energy should be lost by not harvesting it biologically. Lost sun energy is lost opportunity. Photosynthesis is the main process by which Solar Energy is absorbed. It is of course the objective to obtain a higher degree of photosynthesis. Although genetically photosynthesis efficiency is around 1.5% to 2.5%, we can increase leaf index [area of leaf for every square metre of land] by caring for healthy canopies, use of multiple canopy utilising direct and filtered sunrays.

Five Stages in Plant Life

Every plant goes through five stages in its life: Childhood Puberty Youth Maturity and Old age. These stages are of roughly equal duration and external interventions at specific stages are most important. (e.g.

There is no use giving fertilizer dose when the plant has become old and is dying) Generally, plants can be classified as having a seasonal, short duration life span 0 to 130 days], medium life span of 4 – 5 years, or perennial long life span.

For short duration life span, all 5 stages become very critical. For example, if sumptuous roots are not developed in the first 15-20 days [20% of lifespan] no amount of external inputs, thereafter will be useful or effective. Leaves and Branches also show these stages in their life cycles. Yellow, old leaves can only fall and cannot be rejuvenated. Old branches eventually become deadwood.

Medium for Root Zone

Generally, this is soil. However, one can do without it as in the case of hydroponics.

The main purpose of the medium is:

- To give support to the plant and anchoring it by means of shoot root and feeder roots.
- To supply nutrients to the feeder roots.
- To provide moisture to the plant roots.
- To provide good air circulation to the roots.

It is always possible to prepare an ideal soil by human intervention. The soil also supports a whole range of life starting from microorganisms up to earthworms. The presence of these life forms provide essential benefits to the plant roots, in that they convert minerals found in nature into root-absorbable forms.

Plants Manufacture their Own Food

Unlike animals, plants manufacture their own food. By means of photosynthesis, water and carbon dioxide, is converted into sugar: glucose, which is then converted to other forms of sugar, lignin, fats, etc. Plants produce 3 – 4g of dry mass/ square foot of photosynthesis area / per one sunday of 8 – 10 hrs. From this: (i) 1g is used in plant metabolism (ii) 1g is used to build plant body, roots, stem, leaves, etc and (iii)1g is either stored or used for producing fruits. It is very useful and instructive to know, especially about the timings and places of storage of food/ energy and how to tap them at appropriate time. Without the knowledge of this the enzymes and hormones [Gibberilic acid, Indol acetic acid] may lead to a stage, where plants grow with luxurious growth but scarce fruiting.

Limited Water Supply

Assured supply of 1000 litres of water per day is a right of every family. With this as the only external input, prosperity can be built from within the neighbourhood using the Prayog Pariwar methodology.

Implementing Natueco Principle with Prayog Parivar Methodology

a)*Soil Management* - The first step of Natueco Farming is to develop the Nursery Soil using neighbourhood resources. Nursery Soil consists of 50% biomass and 50% activated mineral topsoil by volume. The Biomass forms the organic part and the topsoil forms the inorganic part of the Nursery Soil. The Nursery soil provides support and delivers water and nutrients to the plant in the most efficient manner. To obtain high quality nursery soil, it is most important to build its organic part through biomass addition. The well composted organic part of the nursery soil is called HUMUS which contains ligno proteins. It is a black, light, and easily friable material that can be broken into small fragments or crumbs. It has very good water holding capacity, twice its own weight. Generally, the weight of such material per litre of its volume in fine crumb form is about 400 grams. It has a peculiar black luster & we can see layers of dead colonies of the micro flora especially in well composted (humified) animal dung.

b)*Harvesting the Sun* - It is well known that the entire food chain is directly or indirectly depend upon sun energy, harvested by green plants through a process known as photosynthesis. Although one square metre of land surface usually receive 14,400 K. cal energy per day during bright sunny days, but only a small fraction of direct, total solar insolation is harvested by plants. The philosophy of Natueco culture emphasize on increasing the sun energy harvesting potential of plants, first by understanding the requirement of canopy index establishment, secondly by understanding the principal, that only the mature leaves of the plant are capable of doing optimum harvesting of sun and third by understanding the requirement of matching storage organs.

What is Canopy Index Number: Canopy index number is calculated by accommodating overlapping leaves in multiple of 3 in one sq.ft. area. For example a plant, whose up to 3 overlapping leaves covers an area of 1 sq.ft., assigned Index No.1, with six overlapping leaves in 1 sq.ft., assigned Index No.2, with 9 overlapping leaves in 1 sq.ft. assigned Index No.3, with 12 overlapping leaves in 1 sq.ft., assigned Index No.4 and so on. Different plants have different leaf indices, generally between 5 and 10. Even the same plant can have different leaf indices as per its vigour. Maximise Sun Harvesting - Critical studies by Prayog Parivar revealed many new findings, summerised below:

i. To harvest optimum sunlight, the plant must build a canopy as per the index number of the crop, which is generally between 5 to 10. Thus if a plant's index number of leaves is 5, it means that, to harvest maximum sunlight from one sq.ft of area, this plant must have a canopy of 5 sq.ft. area.

ii. Each plant needs specific area for its maximum growth. By multiplying this area required by the plant, by the index number of leaves, we can calculate the requirement of canopy area e.g. if a plant needs about one half sq.ft. area for its full spread and its leaf index is 5, then for optimum sun harvesting the plant must have a canopy of 2.5 sq.ft. (0.5 x 5).

iii. Each plant must establish its optimum canopy spread at the earliest of its life cycle.

iv. As only mature leaves are capable of doing optimum harvesting of the sun, the young emerging leaves and old dying leaves should be excluded from the point of view of calculation.

v. There should be matching storage organ growth in plants at the time when optimum photosynthesis is taking place in the matured leaves.

Recycling Process

For optimum and continued growth of crops in the fields year after year, it is very much essential that recycling process caries on effectively. Three major components of this recycling process are (a) aerial component such as carbon dioxide, oxygen and nitrogen (b) mineral elements from soil and (c) water. For optimum availability of all the essential nutrients, the fertility of the soil is very important. A good fertile soil maintains its fertility structure, only when the mineral part and the composted part of the soil are equal by volume. In Prayog Parivar's term such soils are called nursery soils. Thus in one cubic foot of nursery soil one half cubic foot soil will be of mineral part and the other one half of the well composted part of dead organic matter. Decomposition of dead organic matter first into compost and then finally into mineral matter and gases is the last part of the food chain of our ecosystem.

(a) Among the aerial components, although one need not worry about carbon dioxide and oxygen, supply of nitrogen is partly ensured by good population of nitrogen fixing bacterial in fertile soil and partly is obtained from decomposition of degrading organic matter.

(b) Entire quantity of mineral elements, need to come from soil reserve. Due care is needed to keep a balance in demand and supply of these nutrients by recycling partly the same crop residue and if needed from neigbourhood resources. Most of the nutrients absorbed by the plants are conserved in their leaves and twigs and only a small fraction goes in fruits and grains. To maintain the balance, the entire quantity of leaf and twigs etc. need to be recycled back and to compensate the loss of a small fraction of nutrients taken away in fruits and grains need to be compensated by small biomass /compost from other sources such as animal dung etc.

(c) The water cycle is the cycle run by the nature, bringing water from the oceans to the lands and the mineral contents from the lands to the oceans. The evaporation of one litre of sea water requires 600 K cal of sun energy. The clouds so generated are transported to the lands by strong winds which eventually come down as rain. All life cycle on land depends on the supply of rain water and the life cycles in the oceans depend on the supply of nutrients carried through the water returning back to the seas or oceans. The Natueco culture cautions that in bringing one litre of rain water about 600 – 800 K cal of sun energy is involved, therefore we should not allow this precious water to be wasted as run-off, before it is fully utilised to optimise biomass production at every place. As per Prayog Parivar's estimate, the water required for producing one Kg biomass (on dry wt basis) is about 6000 gms. Only this much water is used in the production of 1 Kg carbohydrates. The remaining quantity of water supplied to the plant is used only temporarily and released in the environment, creating micro-climate of moisture conditions, favourable for its growth, as well as for the growth of the ecosystem.

Energy Pool and Energy Chain

The Nature's food chain starts with synthesis of carbohydrates in green plants, runs through various macro and micro consumers and ultimately ends into brown mass of decomposed organic matter, before releasing its mineral elements in air and soil. Although, this process is a natural process and runs without the involvement of human but man has accelerated the process to his advantage by bringing more land under cultivation, by using tools and the energy of drought animals. The use of animals in this sun harvesting is very important and must be exploited to its best, otherwise, it will be just wasted. The use of man-made energy tools might have accelerated the process, but has rendered the animal energy useless which is going waste. In Prayog Parivar's vision all natural aspects, which can be incorporated in accelerating the biomass production need to be exploited to its optimum to harvest maximum richness of the nature.

Disease Management

For a vegetable to become truly diseased, several conditions must be present: a susceptible host plant, a pathogenic organism, a good method of distributing the organism, and the proper environment for it to exist, enter the plant, and thrive. When these conditions are met, infection occurs, and a disease agent becomes established. The choice of a proper management tactic must be based on accurate knowledge of the pathogen causing the disease; its life cycle; time of infection; the part of the plant involved; the method of agent distribution; past, present, and future environmental conditions; and certain economic considerations. Effective management techniques include: use of resistant varieties; use of noninfested soil or long rotations; sterilization of soil with steam or chemicals; use of clean seed, either certified or grown in disease-free areas; treatment of seed with heat or chemicals; control of insects and weed hosts; monitoring of weather conditions; use of biological control agents; and proper timing and application of fungicides or nematicides.

Effective management of vegetable diseases requires preventing disease or, if this is not feasible, slowing the spread of disease once it occurs. What can be done to prevent disease outbreaks or reduce the risk of early-season epidemics? Nine procedures and the current estimated percentage of importance of each toward vegetable disease control, have been recommended for many years: rotating crops (30%), spraying when necessary (20%), treating the seed (15%), using clean seed (10%), planting resistant varieties (5%), controlling weeds (5%), aerating the soil properly (5%), draining and fertilizing the soil (5%), and practicing good sanitation (5%). It is unlikely that all diseases of a particular crop can be controlled by simply following these procedures. Nevertheless, the extent of disease and the concomitant costs of controlling them can be significantly reduced by following as many of these procedures as possible. Growers should note that this estimate indicates that spraying is only responsible for 20 percent of disease control. Using the other disease control techniques, which contribute 80 percent of disease control cannot only greatly improve disease control, but also lessen the costs of spray materials and result in better quality crops.

Diagnosis of Disease

The first step in disease management should be accurate diagnosis. It is important to differentiate between infectious diseases (e.g. those caused by fungi, bacteria, viruses, and nematodes that can spread from plant to plant) and noninfectious diseases or disorders (e.g., damage caused by mites and insects, physiological disorders, air pollutants, nutrient imbalances, and herbicide injury). Growers who have a reasonably good understanding of plant diseases, their symptoms, and the infectious and noninfectious disorders that can affect a particular crop, are more likely to make the correct disease control decisions. Numerous fact sheets and bulletins with full-colour illustrations have been developed by Cornell faculty to assist growers in making accurate disease diagnoses. In addition, samples can be sent to the Plant Disease Diagnostics Clinic in Ithaca (607-255-7850).

Disease Management Tactics

Crop Rotation and Tillage: Rotating fields to different crops each year cannot be overemphasized as one of the most important and easily implemented disease control strategies. This practice avoids the buildup of certain plant pathogens in the soil. The longer the rotation, the less likely that an early-season disease outbreak will occur. Because pathogens usually attack members of the same plant family, it is best to avoid planting successive crops belonging to the same family. Choices of unrelated crops to be rotated include beans to sweet corn, leafy vegetables to cucurbits, cucurbits to crucifers, and crucifers to sweet corn. Rotating beans with a grain crop such as barley, oats, rye, wheat, or field corn or with a forage crop is very beneficial for root-rot control. One or two years in a grain crop is often long enough to prevent severe root rot when the field is not heavily infested.

Some soilborne diseases are not readily controlled by rotation. Such diseases are caused by pathogens that produce structures that can withstand the effects of time and nonhost crops. Examples include clubroot of crucifers, Phytophthora blight, and Fusarium wilt of several crops. Other pathogens have such a wide host range that they can survive indefinitely because so many crops and weed species serve as hosts. These pathogens include Sclerotinia, Rhizoctonia, Verticillium and rrot-knot nematodes. Other pathogens are not affected by rotation because they overwinter in southern states, and new inoculum is blown into the area every year. This group includes sweet corn rust and downy mildew of cucurbits.

Many pathogens can overwinter successfully in association with plant debris and are unable to survive once the crop residue decomposes. Destruction of current-season crops can eliminate reservoirs for overlapping plantings. Fall tillage is important because it reduces the amount of inoculum that survives the winter.

Disease-resistant Varieties

The use of disease-resistant varieties is among the most economical and reliable methods of disease control. Although resistant varieties are not available for all diseases of veg

table crops, this option should be used whenever possible. Resistant varieties exist for asparagus, bean, cabbage, cucumber, muskmelon, pea, potato, spinach, sweet corn, and tomato diseases that are important in the northeastern states. The list of disease-resistant varieties will surely increase in the coming years. An advantage of using varieties that are resistant to soilborne disease can be the long-term decline in the pathogen population in the soil when used in combination with adequate crop rotations. Use of varieties resistant to foliar diseases can prevent expenditures and potential environmental consequences of fungicide applications.

Healthy Transplants and Seed

A basic rule for controlling plant diseases is to begin each growing season with healthy seed and transplants. A crop established with infected or infested plant material may contaminate an entire field and remove it from production for many years. Introduction of diseased material as primary inoculum into a field will greatly increase the chance of early-season epidemics, resulting in reduced yields, poorer quality products, and added costs for chemical control. Although diseases may occasionally be introduced via seed from commercial companies, seed companies are the most reliable source of plant material. Saving vegetable seed from the previous year's crop is a certain way of perpetuating a disease from season to season. So too is growing transplants for a friend with seed from his or her "private stock." In recent years, many vegetable operations have diversified, combining growth of bedding plants, maintenance of stock propagating material, and production of local vegetable transplants, often within close proximity.

The potential for spread of vegetable and ornamental plant diseases has been documented with tomato spotted wilt virus and its vector, the western flower thrips. Another example is powdery mildew (two different species) spread to tomato and cucurbits from ornamental plant sources. Although local transplant production allows a more flexible production schedule, proper hardening of plants, and a reduced risk of introducing southern soilborne and foliar diseases, there may be offsetting disadvantages such as the risk of spreading seedborne diseases and other diseases endemic to northern states or harboured within the plastic houses.

Seed treatment is generally important to obtain and hold good stands, especially when soil moisture and temperature conditions are unfavourable during the germination period. Many seed suppliers treat seed before it is sold. If the seed has not been treated by the supplier, the grower can treat it. Seed treatment is especially necessary for beans, beets, onion, carrot, sweet corn, and peas; available materials are discussed under these crops.

The use of healthy transplants and seeds can help growers avoid many costly and environmentally damaging fungicide applications later in the season. Often, once a disease gets started in a field as a result of poor quality transplants or seeds, crop damage or yield loss will result no matter how many rescue treatments are applied.

Other Cultural Practices

Other practices can be followed to make conditions less favourable for development and spread of disease. These include planting after soils have warmed, selecting well-drained areas, using raised beds, reducing plant densities, scheduling overhead irrigation when foliage will dry soon afterwards, controlling weeds, and avoiding root pruning and stem injury by cultivating too deep or too close to plant stems. For example, steps to minimise the severity of white mold for beans would include planting rows in an east-west direction; using wide row spacing to promote drying of soil and reduce moisture in the plant canopy; not planting in shaded areas and small fields surrounded by trees that reduce air drainage; and not planting in fields that drain slowly and have a history of white mold. Steps to minimise white mold from the cabbage crop would also require the removal of ragweed and velvetleaf from the field.

Disease Scouting

Scouting, or systematically monitoring fields for the presence of pests, is an important aspect of any disease management program for vegetables. Recommended scouting procedures are available for many disease pests and are specifically designed to detect diseases when they reach economically damaging levels. Most often, when scouting for diseases, one should make sure to look for symptoms of the disease on a few plants in many areas of the field. Proper diagnosis and recognition of symptoms is important for successful disease scouting.

Weather Monitoring and Pest Forecasting

During the last 15 to 20 years considerable research in plant pathology has been dedicated toward developing mathematical models of plant disease development which can be used to forecast disease outbreaks. Most often these forecasting systems are driven by environmental factors such as temperature, leaf wetness, rainfall, and relative humidity. To most effectively use the forecasting models, the weather conditions near the fields need to be monitored regularly, and then the data need to be run through the computer models. In the past, data collection has been a time consuming and expensive process. Recent innovations in weather data collection, the use of personal computers, and the development of the Internet have allowed weather data collection for disease forecasting models to become affordable. There are two types of data available to farmers - data that is collected on electronic instruments placed in fields and retrieved by phone lines and data which is extrapolated to local conditions using mathematical equations from central collection sites. In order to make use of the first type of data, growers must obtain software for their own computer or subscribe to a service which connects to their weather instrument and interprets the data into disease predictions for them. In order to make use of the second type of data, growers must subscribe to a service which owns the mathematical models for extrapolation. In many cases, the use of the models will result in fungicide application savings which pay for the data acquisition costs.

Among the forecasting programs available are potato and tomato early and late blight, onion programs for Botrytis leaf blight, Alternaria and downy mildew; and white mold of snap beans. In addition, weather monitoring information can be useful for analysis of crop growth and yields.

NEWA makes daily disease forecasts for a number of pests of vegetable crops in New York based on weather data collected in growers' fields. This information is available to NEWA members who pay a fee to join the organisation and support its activities.

Thresholds

Both disease scouting information and weather information require the use of disease thresholds to be useful. Thresholds are levels of disease or disease conducive weather beyond which it is recommended that growers take some action to avoid economic loss from crop disease. Many common crop diseases occurring in New York have threshold information available for them. This information is listed in the disease management tables under each crop.

Biological Control

Research has shown that some fungi and other organisms show activity against specific pest organisms, including both diseases and insects. Some of these organisms have been commercialised and are now available in the marketplace. These products can be quite effective when used properly and offer several advantages over chemical controls, including lessened environmental impact and increased worker safety. Often, biocontrols work more slowly and must be used in combination with other control techniques to be effective. Sometimes fungicides are active against both the pest organism and the biocontrol organism. For this reason, growers should attempt to minimise chemical control applications when trying to make use of biocontrols.

Chemical Control

As the profit margin for successful farming continues to shrink, growers are beginning to look at all production costs in an effort to reduce expenses. Today, as never before, the decision to use chemicals to control diseases can either save a crop from certain economic loss or result in a loss of financial resources. Making the proper decision may depend on the grower's knowledge of the disease in question. With this basic information and by reading labels, the grower can select the appropriate material for treatment. Growers can save money and avoid crop loss by making use of scouting, weather monitoring, disease forecasting and thresholds as described above, before making the decision to apply a chemical disease control agent. Minimising the use of these agents is also a method for growers to practice good environmental stewardship.

Bactericides. In general, there are two types of bactericides: copper compounds and antibiotics. However, products that can induce systemic acquired resistance (SAR) in plants have been shown to affect the development of bacterial diseases (i.e. Actigard). Similarly, some conventional fungicides have shown activity in the suppression of bacterial diseases (i.e. Tanos). All of these products can play a role in reducing the incidence of early- and midseason bacterial epidemics. They are most effective when used in conjunction with cultural practices, including rotating crops, using disease-free seed and transplants, modifying irrigation methods, and regulating activities in a field when bacterial diseases are present. Copper compounds are most effective when disease incidence is low at the time of the initial application and if protection can be maintained during extended periods of weather favourable to disease. Antibiotics serve a similar purpose in certain crops. Because rain helps spread bacterial diseases, an extended period of dry weather will often arrest an epidemic.

Fungicides. Fungicides can be classified as protectants and eradicants. Protectant fungicides act as a chemical barrier to infection by plant pathogenic fungi. They must come into direct contact with a germinating spore or growing mycelium to be effective in preventing subsequent spore germination and infection. They may be used as seed or soil treatments or as foliar sprays. Because these fungicides are not site specific and can control a diverse group of pathogenic fungi, they are used widely for disease control. Additional terms used to describe protectant fungicides are *preventive*, *contact*, and *broad spectrum*. Because protectant fungicides are not absorbed by the plant per se, they do not destroy or burn out existing infections. Once an infection has occurred, a lesion will develop and may produce more spores despite the presence of the protectant fungicide. To be effective, a protectant fungicide must be applied repeatedly during the season and in such a manner as to allow for maximum spray deposition and coverage. Eradicant fungicides have been developed for the control of a limited number of fungi. They are also called systemic fungicides because they are absorbed into the plant and are able to eradicate existing infections. Their main advantage is that they can be applied after infection has occurred and still be effective; because they move systemically within the plant, coverage is less critical; and eradicant fungicides do not need to be applied as often as other fungicides. A major limiting factor of eradicants is that their more specific mode of action may lead to the development of new strains of some pathogens that are resistant to the fungicide. Improper use of these materials can enhance the development of fungicide resistance, such as initiating sprays when disease is well-established. Tank mixing with a protectant fungicide helps maintain the effectiveness of the eradicant.

Fungicides have been arranged by Group Names or Chemical Groups and assigned a Group Code Number by the Fungicide Resistance Action Committee (FRAC), especially to indicate their ability to develop resistance to populations of fungi. The vegetable fungicides that are "at risk" include the following groups:

Group 1	benzimidazoles (ex. thiabendazole, Mertect, and the previous product benomyl, Benlate) and thiophanates (ex. thiophanate-methyl, Topsin M);
Group 2	dicarboximides (ex. iprodione, Rovral and the previously registered product vinclozolin, Ronilan);
Group 3	DeMethylation Inhibitors (DMIs) *imidazoles* (ex. triflumizole, Procure and *triazoles* (ex. myclobutanil, Rally; propiconazole, Quilt and Tilt; and the previously registered product triadimefon, Bayleton);
Group 4	phenylamides, *acylalanines* (metalaxyl, Ridomil and mefenoxam, Ridomil Gold or Ultra Flourish);
Group 7	carboxamides (ex. flutolanil, MonCoat and † Moncut; boscalid, Endura)
Group 9	anilinopyrimidines (ex. cyprodinil, Vanguard and pyrimethanil, Scala);
Group 11	Quinone outside Inhibitors (QoI) (ex. azoxystrobin, Quadris and Heritage; pyraclostrobin, Headline and Cabrio; trifloxystrobin, Flint and Gem
Group 12	phenylpyrroles (ex. fludioxonil, Maxim);
Group 13	quinolines, (ex. Quintec, quinoxyfen);
Group 14	aromatichydrocarbons (ex. quintozene (PCNB), Blocker)
Group 15	cinnamic acids (ex. dimethomorph, Acrobat or Forum);
Group 17	hydroxyanilides (ex. fenhexamid, Decree);
Group 21	Quinone inside Inhibitors (Qil), cyanoimidazole (ex. cyazofamid, Ranman);
Group 22	benzamides (zoxamide, * Gavel);
Group 25	glucopyranosyl antibiotic (ex. Streptomycin) (ex. streptomycin sulfate, Agri-mycin 17, and Agricultural streptomycin, and Firewall
Group 27	cyanoacetamideoximes (cymoxanil, Curzate);
Group 28	carbamates (ex. propamocarb, * Previcur Flex).

Do not apply fungicides in these groups exclusively in a disease control program. Fungicides in these groups should be rotated with broad spectrum (multi-site Groups M1, M3, M5 given below), or used in combination with another group of fungicides to delay the development of resistant strains of fungi.

Resistance problems is generally not recognised for the follow groups:

Group 29	2,6-dinitroanilines (ex. fluazinam, Omega)
Group 30	organo tin compounds, tri phenyl tin compounds (ex. fentin hydroxide, Super Tin)
Group 33	phosphonates, ethyl phosphonates (ex. fosetyl-Al, Aliette), and phosphorous (ex. phosphorous acid, ProPhyt and Phostrol);
Group 40	*carboxylic acid amides* (ex. dimethomorph, Forum; mandipropamid, Revus);
Group 43	*pyridinylmethly-benzamides* (ex. fluopicolide, *Presidio;
Group M	(multi-site activity):
Group M1	inorganics (ex. coppers, Kocide, Champion, or OLF; and sulfur, Microthiol Disperss);
Group M2	*inorganics* (ex. sulfur, Microthiol Disperss or OLF);
Group M3	*dithiocarbamates* (ex. mancozeb, Dithane, Manzate, Penncozeb; metiram, Polyram; thiram; ziram, Ziram 76DF; Group 4 *phthalimides,* captan (seed treatment only);
Group M4	*phthalimides,* captan (seed treatment only);
Group M5	*chloronitriles* (ex. chlorothalonil, Bravo, Echo, Equus);
Group 21	(host plant defence induction) *benzo-thiadiazole* (ex. acibenzolar-S-methyl, Actigard)
Group B	Biologicals (ex. fungal and bacterial species, Contans, Serenade, Sonata, and T-22).

Not Classified (ex. mineral oils, organic oils, and potassium bicarbonate)

Do not use "at risk" fungicides as a rescue treatment for disease control. "At risk" fungicides should be used in a full season disease control program or not at all. Applying "at risk" fungicides only after a disease is present in a field, increases the chances for the development of resistant populations of plant pathogenic fungi.

The need for a fungicide spray depends on several factors, including plant stress (caused by fruit maturity or lack of good fertility); weather conditions conducive to fungal spore germination and infection (moisture and temperature); stage of crop development; levels of host resistance; and levels of pathogen inoculum.

Nematicides and Fumigants: These chemicals reduce populations of nematodes and soilborne fungi. Fumigants are usually applied before the crop is planted, whereas most of the non-fumigant nematicides are applied shortly before or during planting. They are most effective when used in nation with cultural control strategies such as crop rotations and resistant varieties. However, the cost-benefit of using such fumigant nematicides varies greatly from crop to crop and their application may be restricted to only custom-applicators.

A number of factors have a pronounced effect on the success or failure of soil fumigation. Six are given below.

Soil Preparation Prior to Fumigation: Soil should be plowed deeply (ten inches or more) in order to incorporate previous crop debris as thoroughly as possible and to prevent the turn ng up of nonfumigated soil during fitting in the spring. This should be followed by disking or any other means of fitting which will leave the soil in seedbed condition. Clods and poorly incorporated debris will provide "chimneys" through which fumigant can escape prematurely from the soil.

***Soil Moisture*:** The soil should be neither too wet nor too dry. A good rule of thumb is that moisture content is most favourable when soil will just "ball" in one's hand when pressure is applied. If soil is excessively dry and irrigation is available, moisture supplementation before fumigation is recommended.

***Soil Temperature*:** The optimal temperature for most fumigants is 50° to 70°F. At warmer temperatures, fumigants dissipate thoroughly and rapidly, nematode larvae (which are easier to kill than eggs) have emerged, and all nematode stages can be more effectively controlled.

***Crop Debris*:** Undecomposed residues from previous crops prevent distribution of fumigant through the soil, irreversibly absorb fumigant, interfere with application equipment, prevent proper sealing of the soil surface, and protect nematodes and nematode eggs from fumigant action. Rake, burn, or deeply incorporate debris prior to fumigation.

***Sealing of Soil Surface*:** It is essential that fumigated soil be thoroughly sealed as soon after application as possible. This can be achieved by means of equipment such as a cultipacker, chain harrow or float, or by means of spray irrigation or plastic sheets.

***Interval between Fumigation and Planting*:** Under average conditions, with a soil temperature of ± 50°F, a minimum of three weeks is regarded as necessary between fumigation and planting to prevent phytotoxicity to potatoes.

The following multipurpose soil fumigants should be used to provide disease and nematode control;

Chloropicrin—50 gal/A

or

metam sodium (Vapam HL)— 37.5- 75 gal/A

or

metam potassium (K-Pam HL)—30-60 gal/A

or

methyl bromide (*Terr-O-Gas 67 or 98,* MBC-33)—225-350 lb/A

or

dichloropropene + chlorophicrin (*Telone C-17) [10.8-17.1 gal/A mineral soil; 27.4-30 gal/A muck soil]; *†Telone C-35 [13-20.5 gal/A mineral soil; 33-36 gal/A muck soil].

A plastic film seal is needed when methyl bromide is used, and it will increase the efficacy of all the above treatments. Other soil fumigants are injected to a depth of 6 to 8 inches. Immediately after application, soil should be dragged, rolled, or cultipacked to delay loss of fumigant. Metam-sodium is water soluble and can be injected and applied via field is irrigated (apply an acre inch of water). Rinse the irrigation system with clean water only long enough to clear the system. Too much rinsing or heavy rainfall within 24 hours will reduce the efficacy of the treatment.

At least 2 to 3 weeks should intervene between the application of most soil fumigants and the time a crop is planted. Methyl bromide requires only 2 to 7 days of waiting after removal of the tarp.

One week after application, work soil to a depth of several inches so that gasses may escape. Severe injury killing of sensitive plants may occur if the fumigant has not sufficiently dissipated.

To determine if it is safe to plant into fumigated soil, collect a soil sample from the treated field (do not go below the treated depth). Place the sample in a glass jar with a screw top lid. Firmly press numerous seeds of a small seeded vegetable crop (lettuce, radish, etc.) on top of the soil and tighten the lid securely. Repeat the process in another jar with nonfumigated soil to serve as a check. Observe the jars within 1 to 2 days. If seeds have germinated, it is safe to pant in the field. If seeds have not germinated in the fumigated sample and have germinated in the nontreated sample, then the field is not safe to plant. Rework the field and repeat the process in a few days.

Owing to a reduction in nitrifying bacteria by the fumigants, at least 50% of the nitrogen in the initial fertilizer application should be in the nitrate form.

Permitted Pesticides in Organic Farming

The management of insect pests rarely relies on a single control tactic; usually, a variety of tactics are integrated to maintain pests at acceptable levels and minimise the chance that insects will adapt to any one management tactic. The goal is not to eliminate all pests, but rather, to reduce pest populations to tolerable levels. Some pests are tolerable and essential, so that their natural enemies remain in the crop. The management tactics used against insect pests include pest resistant or tolerant varieties, and cultural, physical, mechanical, biological, and chemical controls.

Integrated pest management requires an understanding of the ecology of the cropping system including that of the pests and their environment. For example, temperature is the primary factor determining the rate at which insects develop; higher temperatures increase the rate of development.

Temperature can be important when determining the frequency of insecticide applications. In sweet corn for example, the interval between insecticide applications for the European corn borer can be increased when the weather is cool because eggs are not hatching as rapidly and the rate of larval development is slower; with warmer temperatures the situation is reversed. Degree-day models can aid in determining how fast insects are developing and the timing of applications. Weather conditions can also influence insect pest populations. For example, wind-driven rain is a major source of mortality to nymphs of the potato leafhopper on potatoes and thrips on onions and frame leaves of cabbage.

Economic Threshold and Sampling: The decision to use an insecticide, or take some other action, against an insect infestation requires an understanding of the level of damage or insect infestation a crop can tolerate without an unacceptable economic loss. The level of infestation or damage at which some action must be taken to prevent an economic loss is referred to as the "action threshold." Action thresholds are available for many vegetable crops and should serve as a guide only.

Thresholds should be adjusted based on market, environmental conditions, variety, etc. To estimate the severity of pest infestations, the crop must be sampled. Sampling may involve examining plants and recording the number of pests or the amount of damage observed, or traps may be used to capture the pest species to estimate pest abundance. Sampling is conducted at regular intervals throughout the season or during critical stages of crop growth.

Weather and predicting insect occurrence: Some insect pest activity can be predicted by using degree days - a method of summarising accumulated heat units.

Management Options

Pest-resistant Crops: An important management option for the control of insect pests is the use of crop varieties that are resistant or tolerant. A resistant variety may be less preferred by the insect pest, adversely affect its development and survival, or the plant may tolerate the damage without an economic loss in yield or quality. For example, varieties of cucurbits (squash, cucumbers, melons) that have lower concentrations of feeding stimulants (cucurbitacins) are less preferred by cucumber beetles. Sweet corn varieties with tight husks are less likely to be infested by corn earworm, and some varieties are resistant to Stewart's wilt, a bacterial disease of sweet corn vectored by the corn flea beetle. Bt sweet corn varieties can be considered resistant varieties developed through modern biotechnology. In the case of cabbage, the only reliable method of controlling onion thrips is through the use of tolerant varieties. Advantages of pest-resistant crop varieties include ease of use; compatibility with other integrated pest management tactics; low cost; cumulative impact on the pest (each subsequent generation of the pest is further reduced); and reduced negative impact on the environment.

Cultural Control

There are many agricultural practices that make the environment less favourable to insect pests. Crop rotation, for example, is recommended for management of Colorado potato beetle. The beetles overwinter in or near potato fields, and they require potato or related plants for food when they emerge in the spring. Planting potatoes well away from the previous year's crop prevents access to needed food, and the nonflying beetles will starve. Selection of the planting site may also affect the severity of insect infestations. Cabbage planted near wheat or oats is more likely to be infested by onion thrips that disperse from the maturing grain crops.

Trap crops are planted to attract and hold insect pests where they can be managed more efficiently and prevent or reduce their movement onto crops. Early planted potatoes can act as a trap crop for Colorado potato beetles emerging in the spring. Since the early potatoes are the only food source available, the beetles will congregate on these plants where they can more easily be controlled. Adjusting the timing of planting or harvesting is another cultural control technique. Earlier planted sweet corn is less likely to be infested by corn earworm and fall armyworm, pests that typically arrive mid to late in the season.

Physical and Mechanical Control

The use of physical barriers such as row covers or trenches prevents insects from reaching the crop. Row covers can help prevent early-season damage to cucurbits by cucumber beetles, and plastic-lined trenches are effective in trapping dispersing Colorado potato beetles in the spring and fall. Other methods include handpicking of pests, sticky boards or tapes for control of flying insects in greenhouses, and various trapping techniques.

Biological Control

Biological control is defined as the reduction of pest populations by natural enemies (predators, parasitoids, and pathogens). Predators, such as lady beetles and lacewings, are free-living species that consume a large number of prey during their lifetime. Parasitoids (certain wasps and flies) are species whose immature stage develops on or within a single insect host, ultimately killing the host. Pathogens are disease-causing organisms including bacteria, fungi, and viruses. Pathogens kill or debilitate their host and are relatively specific to certain insect groups.

Natural enemies can be either generalists, which utilise many different species of insects as food, or specialists, which are fairly selective in their prey. Lady beetles are considered generalists because they feed on many types of prey; whereas, species of *Trichogramma* are microscopic wasps specialised to parasitize eggs of moths. There are three broad and somewhat overlapping types of biological control: classical, augmentative, and conservational.

The conservation of natural enemies is probably the most important and readily available biological control practice available. Natural enemies occur in all vegetable production systems. They are adapted to the local environment and to the target pest, and their conservation is generally simple and cost-effective. With relatively little effort, the activity of these natural enemies can be observed. Lacewings, lady beetles, hover fly larvae, and parasitized aphid mummies are almost always present in aphid colonies. Fungus-infected adult flies are often common following periods of high humidity. These natural controls are important and need to be conserved when making pest management decisions. In many instances, the importance of natural enemies has not been adequately studied or does not become apparent until insecticide use is stopped or reduced. Often the best we can do is to recognise that these factors are present and minimise negative impacts on them. If an insecticide is needed, every effort should be made to use a selective material in a selective manner.

Natural enemies are generally more adversely affected by chemical insecticides than the target pest. Because predators and parasitoids must search for their prey, they generally are very mobile and spend a considerable amount of time moving across plants. This increases the likelihood that they will contact the insecticide. When an insecticide is applied, ideally only the target pest(s) should be affected. The goal is to maximise pest mortality while minimising harm to natural enemies.

The following factors influence the impact of insecticide applications on natural enemies:

Spectrum of Activity: Broad-spectrum insecticides will more adversely affect natural enemies than materials that are selective for specific insect species or life stages.

Residual (half-life) Activity of Insecticide: Insecticides that remain toxic to pests for a long time and stay on the treated surface will have a similar effect on natural enemies.

Coverage and Formulation of Insecticide: Full coverage sprays will generally have a greater impact on natural enemies than directed sprays, systemics, or bait formulations. Spot or edge treatments directed at localised pest infestations will be less detrimental than those applied to the entire field.

Dosage and Frequency of Application: Higher rates and repeated applications will have a more detrimental impact on natural enemies.

Susceptibility of Natural Enemy to Pesticides: Some natural enemies are inherently more resistant to insecticides than others and some populations of natural enemies have been selected to possess higher levels of resistance. Applications of some fungicides can also reduce the incidence of fungal diseases of insects. Fungicides and herbicides may also have lethal and sublethal effects on natural enemies.

In many instances, the complex of natural enemies associated with an insect pest may be inadequate. This is especially evident when an insect pest is accidentally introduced into a new geographic area without its associated natural enemies. These introduced pests (e.g. European corn borer, imported cabbageworm, carrot rust fly, and diamondback moth) comprise about 40 percent of the insect pests in the United States. Classical biological control is the practice of importing and releasing for establishment (following a quarantine process) natural enemies to control an introduced pest. Follow-up studies are conducted to determine if the natural enemy was successfully established at the release site and to assess its long-term benefit.

Augmentation is the supplemental release of natural enemies. Relatively few natural enemies may be released at a critical time of the season or literally millions may be released. For example, the recommended release rates for *Trichogramma ostriniae* in sweet corn is 30,000 per acre. In contrast entomopathogenic nematodes are released at rates of millions and even billions per acre for control of certain soil-dwelling insect pests. Augmentation may also involve modifying the cropping habitat to favour natural enemies. Many adult parasitoids and predators benefit from sources of nectar and the protection provided by refuges such as hedgerows, cover crops, and weedy borders. The modification of surrounding habitats must, however, be done carefully. Refuges may benefit some pests, but may also be hosts for certain plant diseases, especially plant viruses that could be vectored by insect pests.

Chemical Control

When other management tactics are not available or fail to keep pest infestations from causing economic damage, insecticides are needed. The decision to use an insecticide should be based on an action threshold. A field should be scouted regularly to determine if it is at the action threshold. When an insecticide is required, the choice of material, formulation, and rate and method of application will vary depending on several factors, including the pest(s) to be controlled and the stage of crop development.

Insecticides fall into several chemical classes including: chlorinated hydrocarbons (*methoxychlor*), organophosphates (*malathion* and *terbufos*), carbamates (*methomyl*), synthetic pyrethroids (*esfenvalerate* and *permethrin*), and microbials (*Bacillus thuringiensis*). More recently several new insectocodeshave been developed and these include neonicotonoids (*imidacloprid*), avermectins (*emamectin-benzoate*), naturalyte (*spinosad*) and others. In addition to these common classes of insecticides, there are insecticides derived from plants (*rotenone*) and inorganic materials (*cryolite*). They are also classified by mode of action: physical poisons, protoplasmic poisons, metabolic inhibitors, and nerve poisons. Physical poisons kill by an action such as suffocation and include certain oils. Protoplasmic poisons kill by destroying proteins. Metabolic inhibitors interfere with processes such as respiration or digestion. Nerve poisons, the largest group of insecticides, interfere with the normal functions of the nervous system.

Insecticides are formulated into emulsifiable liquid concentrates, flowable liquids, and dry concentrates such as wettable powders, soluble powders, dusts, and granules. A sticker-spreader should be added to spray mixtures for use on some cole crops, cucumbers, onions, and tomatoes, so the mixture will spread more evenly over the plant surfaces. Insecticides will be effective only if the insect consumes or comes in contact with them; good coverage is thus essential for efficient pest control. Always use enough water and adequate pressure to cover plants completely, especially if insects are concealed (e.g., thrips on onions). Water requirements may vary from five to 150 gallons per acre, depending on the equipment, chemical, crop, and plant size. Use the correct application equipment to assure adequate coverage. Drop nozzles or airblast sprayers may be essential to get the insecticide into the crop canopy and to the pest. Several factors determine the efficacy of an insecticide application, including use of the correct insecticide; proper timing of the application; and good coverage of the crop. Some insecticides are more effective than others for a particular pest, and certain insecticides must be applied at specific times. For example, to be most effective *Bacillus thuringiensis* (Bt) products must be applied early in the pest's development when it is more susceptible to the toxin. Coverage is also important. Aphids, for example, are frequently most abundant on the undersides of leaves. If the insecticide is applied only to the tops of leaves, aphid control will not be adequate. Thorough coverage is also important for stomach poisons, such as the Bt products, which must be ingested by the pest.

Managing Resistance

A major obstacle to the continued, effective use of insecticides is the development of resistance. Insect pest populations develop resistance to insecticides through genetic selection. Some individuals in a population have genetic traits that allow them to survive an insecticide application. Some of the progeny of these surviving individuals will possess these traits. When additional applications of insecticides are made, more of the resistant individuals survive. Over time and in the absence of appropriate resistance management, the pest population may consist mostly of resistant individuals. At this point, control may be difficult or impossible. Increasing the frequency of the applications or amounts of insecticide will not improve control.

Switching to different or new insecticides within the same class or even different classes may not help because pests resistant to one insecticide may also be resistant or rapidly develop resistance to others. This is referred to as cross-resistance. For example, some insect pests that are resistant to chlorinated hydrocarbons quickly develop resistance to pyrethroids. Repeated applications to control a target pest may lead other nontarget species subjected to these same applications to develop resistance.

Many insect populations now possess some level of resistance to insecticides. A partial list includes some populations of green peach aphid resistant to *endosulfan*, and pyrethroids, and some populations of the diamondback moth are resistant to pyrethroids, *methomyl*, and Bt's. The Colorado potato beetle is resistant to several insecticides. The following steps are recommended to minimise the development of resistance to insecticides: apply insecticides only when needed; follow recommended action thresholds where available; use the minimum recommended rate that provides control; rotate insecticides that are in different chemical classes and have different modes of action during the season. Ideally, this rotation should be done across pest generations, that is, each generation should be treated with a different class of insecticide. Also, use insecticide synergists if available and effective.

An insect may be resistant to one insecticide in a particular class (e.g. pyrethroid) but be susceptible to another insecticide in a different class (e.g. carbamate). Therefore it is important to know the class of the insecticide you choose.

Chapter 6

Weed Management

Weeds reduce yield and quality of vegetables by competing directly for light, nutrients, and water and by interfering with harvest operations. Early-season competition is most critical and a major emphasis on control should be made in this period. Weed control requires good management practices in all phases of production. Because there are so many different kinds of weeds, they cannot be managed by a single method. The first step in managing weeds is identification. Some of the common problem weeds in New York State are highlighted below.

Problem Weeds in Vegetable Production

Galinsoga

Because of its biology and its tolerance to vegetable herbicides, galinsoga may quickly become a major weed once it is introduced into a vegetable field. Galinsoga is not sensitive to day length and, as a consequence, begins to flower and produce seed when it has about five or six pairs of leaves. It continues to form seed until it is killed by frost. Fresh seed drops onto the soil surface and soon sprouts because there is little or no dormancy. The new seedlings begin to flower at the fifth or sixth node, and the cycle is repeated. Three to five generations per season are often observed in Ithaca, New York. Cultivation is only partially helpful because Galinsoga roots quickly and easily re-establish themselves from cut stems and uprooted plants unless conditions are very dry for several days after cultivation.

Velvetleaf

This robust, annual weed is increasing rapidly in upstate New York areas. It often escapes in fields where preemergence herbicides are used without mechanical cultivation. It has fairly large seeds that last many years in the soil and are not destroyed when fed to cattle.

Because of their size, seeds can germinate anywhere in the top several inches of soil. Seedlings emerge from a range of depths, appearing over a period of many weeks, and most surface-applied herbicides used at planting are relatively ineffective on seedlings that emerge late. Unfortunately, even late seedlings can produce mature seeds before frost.

Nightshade

Nightshade, a warm-season summer-annual weed, is becoming a problem in New York. Eastern black nightshade (*Solanum ptycanthum*) is the most common and widespread species, but hairy nightshade (*Solanum sarracoides*) is predominant in some areas. These weeds are particularly problematic in tomato, potato, snap bean, and dry bean fields. Few herbicides currently registered for vegetable crops effectively control nightshade. Therefore, to stop an infestation, it is important to identify the weed and eradicate it before the plants produce seeds. *AAtrex*, used in sweet corn, will control nightshade, and thus corn can be used as a rotational crop to reduce control problems in other vegetables. The acetamides suppress nightshades. Additional research will be required, however, to determine the effectiveness of these herbicides in controlling the different nightshade species.

Quackgrass

This common perennial problem grass is controlled most efficiently by a combination of chemicals and tillage. Check specific crop recommendations for control.

Nutsedge (nutgrass)

Tubers do not sprout until the soil is warm. In most fields, weeds such as lambs quarters, mustard, ragweed, and quackgrass emerge two or three weeks earlier. Nutsedge grows vegetatively until midsummer, forming the first tubers as the days shorten slightly in July. As days shorten more sharply in August and September, tuber formation is greatly accelerated. In the fall, even small plants form tubers.

In many vegetable crops, even temporary control is difficult. Furthermore, reducing the severity of a heavy infestation or eliminating nutsedge from a field is difficult because tubers may last four or more years after they have formed. Fortunately, nutsedge is sensitive to dense shade, and successful control measures need to capitalize on this characteristic. For example, when planted early at close spacing, most pumpkins and squash provide the shade needed. For dry and snap beans, potatoes, and sweet corn, nutsedge can be controlled by selective herbicides. Some general suggestions follow: use heavy crop shading if possible; plant crops at close spacings; cultivate between rows until crop foliage takes over; plant and harvest early on fields for which selective chemicals are not available. Fall tillage and nonselective chemicals can then be used. When selective chemicals are available, delay planting and treatment until tubers have sprouted. Herbicides do not damage dormant tubers. Both cultural practices and herbicides are needed; neither can do the job alone.

Perennial Broadleaf Weeds

Perennial broadleaf weeds such as bindweed, Canada thistle, horse-nettle, and milkweed are not easily controlled while vegetable crops are growing. Cultivation alone is only partially adequate, and effective herbicides are very damaging to crops.

Rotation and time of planting should be scheduled, so infested areas are free of crops during late summer and early fall. Suitable herbicides are available, and at this time, perennials are fairly susceptible because, as days shorten, they move food from their leaves downward into overwintering parts in preparation for spring growth. Herbicides applied at this time also are moved downward, causing severe damage. Vigorous green leaves must be present at the time of herbicide spraying. If time permits, the fields should be worked a week or so after spraying, and any areas of regrowth should be given a second application. Severe infestations often require two or more seasons for adequate control.

Hormone-type herbicides such as 2,4-D and Clarity are particularly effective on bindweed and Canada thistle, whereas horsenettle and milkweed require Roundup. Better control of perennial broadleaf weeds is achieved with combinations of herbicides applied after harvest. Using combinations allows growers to reduce the rates of both herbicides while increasing the effectiveness of weed control when compared to either herbicide alone.

Weeds in and Around Greenhouses

Weeds in and around greenhouses and other plant-growing structures are a nuisance as well as a source of insects, diseases, and rodents. Perennial weeds can spread to an entire farm if transplants are infested. Although it is tempting to try to control weeds with chemicals while the crop still occupies the greenhouse, this is extremely dangerous. It is preferable to control weeds with chemicals while the structure is empty, generally in the summer and fall; next, as a preventive measure, lay heavy (six mils or thicker) black plastic over the soil before starting the next crop. Frequent mowing of surrounding areas favours sod and reduces perennials that might harbor pests. Treating a narrow strip next to the walls with Roundup or *Paraquat makes mowing easier.

Caution: Do not permit herbicide sprays to contaminate the upper surface of the plastic. They remain there indefinitely and can damage plant roots or foliage. Do not use hormone sprays such as 2,4-D or Banvel anywhere near a greenhouse. They have a long residual period, toxic vapors, and can contaminate equipment.

Weed Maps

A weed map illustrates problem areas and provides information for future control decisions. When weed maps are kept over a period of years for a given field, changes in weed locations and populations can be noted and decisions about control can be adjusted accordingly. Growers can target areas of severe infestations for specific control practices, rather than treat a larger area needlessly or fail to control problems at all.

Weeds or weed species may be unevenly distributed over a field. Localised areas of severe weed infestations or atypical conditions, such as poorly drained areas, high spots, and field edges, may be recorded on a weed map.

Two weed maps should be prepared during the season. The first should be completed soon after planting to evaluate the success of the current season program. The second should be made at or near harvest to help predict weed control practices that will be necessary for next year.

Weed species: If the species is unknown, at the very least distinguish annuals from perennials, and broadleaves from grasses and yellow nutsedge.

Weed size: The following ratings can be used:

White Sprouts = seeds are just germinating/emerging

Tiny = weeds show only cotyledons or first true leaf

Small = weeds are less than 1" tall (less than the diametre of a quarter)

Large = weeds are more than 1" tall (more than the diametre of a quarter)

Density of each species: Use the following code:

0 = None

1 = Scattered, few weeds

2 = Slight, 1 weed per 6 row feet

3 = Moderate, 1 weed per 3 row feet

4 = Severe, more than 1 weed per 3 row feet

Distribution of weeds: Rate as follows:

Spotty = found in a few places around the field

Local = found in a small portion of the field

General = found throughout the field.

Management Options

Prevention

The higher the weed seed populations in the soil, the more difficult it is to achieve good weed control. Therefore, weeds should be prevented from seeding in and around fields whenever possible. Weeds should be destroyed immediately after harvest. Keeping weeds out of fence rows and off ditch banks can also help.

Crop Rotation

A crop rotation is a planned sequence of crops grown in succession in the same field over an extended time. Crop rotations prevent continuous and uniform management practices from being employed that select for weed species best adapted to a particular system. Crop sequence in a rotation dictates: herbicide use, tillage type, timing of tillage relative to weed and crop emergence, and harvest date relative to weed and crop maturity. The more dissimilar that the crops and their management practices are in rotation, the less opportunity a weed species has to become dominant over several years. Crop variables which will effect yield include: planting/maturation dates, growth habit, competitive ability, crop-specific cultural practices (e.g. number and timing of cultivations), and fertility management. One can say that the most dominant factor influencing weed species composition over time is crop rotation, and that rotation is the most effective means available for suppressing weeds. Traditional rotations include:

- a sod crop (legume or grass) which provides competition from densely-sown species, improves soil organic matter and does not disturb the soil (increasing seed decay time).

- a "cereal" smother crop (provides competition for resources, allelopathy).
- a legume seed for forage crop (serves as a nitrogen resource).
- a root or row crop — also called "cleaning" crops because soils are disturbed and the crops are usually cultivated, disrupting weed growth, and deplete weed seed reserves in the soil by stimulating germination.

Rotations with these components are superior to monoculture systems in maintenance of organic matter, topsoil conservation, and higher yields.

Cultivation

The high cost of skilled and hand labour has caused growers to rely increasingly on chemicals rather than on cultivation and hand hoeing or pulling. Unfortunately, chemicals alone cannot do an efficient, satisfactory job. A few weeds always escape, and new, chemically-resistant weed species may begin to increase. One early, shallow cultivation often solves both of these problems and benefits crop growth and development.

Stale Seedbed Technique

In several vegetables, such as the field-seeded vine crops, weed control is difficult because there are no fully satisfactory selective herbicides available. A stale-seedbed technique often helps in these situations. This procedure involves preparing the seedbed in a normal manner, but perhaps a little earlier than usual; delaying planting for two to three weeks; then spraying with a herbicide to kill emerged weeds; and planting without reworking the seedbed. In soils that tend to bake or crust severely, the stale seedbed method is often not feasible. In good tilth, however, modern seeders work satisfactorily even if rain occurs after seedbed preparation and before seeding. *Gramoxone and Roundup are safe for vine crops with a stale seedbed.

Mulches

Black plastic is the most widely used mulch and can be a good alternative to herbicides in some cases. Black plastic stops most weeds from germinating but some, like nutsedge, can poke through the mulch and become a problem. Clear plastic results in warmer soil temperatures than black which can speed harvest but it does create an ideal situation for weeds and herbicides must be used under clear plastic to avoid competition to the crop. IRT or infrared-transmitting plastic is the third type of plastic mulch used. It does a better job than even black plastic in eliminating weeds and warms the soil better than black but not as well as clear.

Chemical

Herbicides used in Combination or Alone: Because growers are often too busy to accomplish all necessary jobs, they may wish to combine herbicides with other pesticides or fertilizers. Most such combinations are not recommended, however, because they frequently lead to inferior control or injury to crops. It is particularly dangerous to combine liquid fertilizers and postemergent weed killers. Nevertheless, certain combinations of herbicides are highly desirable. In these combinations, each chemical is used at a relatively low dosage, and a larger variety of weeds can be controlled than with single chemicals.

The vegetable being grown dictates the choice of herbicide for a particular field. In many situations, the target weeds cannot be controlled chemically. If a different crop is grown, however, another herbicide may control the weeds readily. For example, ragweed is not harmed by most herbicides that can be used on cabbage. Conversely, several herbicides used in corn and beans are harmless to the crop but toxic to ragweed. This principle applies to many crops and weeds. Thus, crop rotation is a simple, effective means of obtaining good, long-term weed control.

Herbicide Incorporation: Herbicides require incorporation for the following reasons: to reduce loss of efficacy through photodegradation, volatilisation, or evaporation; to activate the herbicide through contact with soil moisture; to place the herbicide in the area where it will have the greatest effect; and to prevent drift.

Herbicides have different physical properties (eg. solubility and volatility) that effect their movement and persistence in the soil. Depth of incorporation is a function of these physical properties. Those herbicides that are volatile tend to require deeper incorporation while those that need moisture for activation tend to require shallow incorporation. Combining herbicides that have different requirements will result in loss of efficacy of one of the herbicides and should be considered when deciding to combine applications.

Two important herbicides that should be incorporated for best results are Eptam and Treflan. Eptam is incorporated (two to three inches deep) to reduce losses from volatility or evaporation. Treflan, on the other hand, is incorporated to ensure its presence in moist soil in the vicinity of sprouting weed seeds. Treflan will fail in dry soil because its solubility is exceptionally low. In contrast, Eptam's volatility is reduced in dry soil, and its performance is improved in moist soil. Granular formulations of Eptam are somewhat less volatile and sometimes more effective than liquid sprays. Treflan is much more stable than Eptam and need not be incorporated for several hours or may be incorporated with irrigation.

The acetamide herbicides do not require incorporation if precipitation is adequate or if irrigation is possible. However, surface applications during drought periods will perform poorly and in this situation, incorporation is recommended. These herbicides require shallow incorporation (called 'surface blending' on some product labels) in the top one to two inches of soil. If these herbicides are incorporated too deeply, efficacy is lost due to dilution in the soil profile.

Chemicals to be incorporated should not be applied to rough, cloddy soil because the herbicide will be distributed unevenly as the clods break up. Several tools are used for incorporation, but all are not equally satisfactory.

The double disk, a popular tool for incorporation, is highly effective in soil preparation, but large, modern disks may give poor results. Many have a preset angle, and the operator can control only depth and tractor speed. Too often, the preset angle results in an average which in practice is inadequate in some fields and excessive in others. An excessive angle throws too much treated soil to the outer edges of the disked area and leaves too little in the centre. An inadequate angle causes poor mixing. Although the latter problem can be overcome by cross disking and the soil may be thoroughly mixed with excessive angle and cross disking, the uneven lateral distribution of the treated soil will cause irregular or streaked patterns of poor weed control and occasional crop injury.

These problems can be avoided by the use of a spring-tooth harrow. In fields relatively loose and free of clods, the old-fashioned corn or potato weeder (sometimes called a finger weeder) also gives excellent results. These tools produce shallow incorporation and require a smaller amount of herbicide and much less power. The spike-tooth harrow, rotary hoe, and cultimulcher have been unsatisfactory for incorporating herbicides.

Water Volume and Nozzle Pressure: For most herbicides, the water used during spraying serves no purpose other than to properly distribute the chemical. If appropriate equipment were available, herbicide concentrates could be applied without water. From a practical standpoint, a rate of 20 to 40 gallons per acre handles almost all herbicides well. Less water may cause excessive nozzle plugging or result in spray droplets so small that they are easily carried away by wind. Higher rates of water require too much time to refill the sprayer.

About 40 psi pressure at the nozzle is preferred for most herbicides. Lower pressure may give an uneven spray pattern, and higher pressure in regular flat-fan nozzles will cause many fine, fog-like droplets, which often result in excessive volatility and serious drift. Furthermore, small droplets of postemergence sprays cause heavier than normal herbicide deposit on plant leaves, which often gives excellent weed control, but can cause crop damage. Weeds should be sprayed when they are less than one inch tall. At this stage, high pressure is unnecessary. A few growers have tried to reduce chemical dosage by increasing sprayer pressure, sometimes successfully. However, until research on this can be evaluated, normal rates and pressures at the appropriate time are the best practice.

Managing Herbicide Resistance and Persistance

Triazine-resistant weeds, Triazine-resistant weeds: Since the early 1970s many weeds have developed resistance, first to the triazine herbicides and more recently to paraquat and the sulfonyl urea and imidazolinone herbicides. Triazine-resistant common lambsquarters was discovered in New York in 1978 and now is ubiquitous in field corn production areas. Triazine-resistant redroot pigweed has increased and now may be found in several hundreds of thousands of acres. Recently there have been at least two documented cases where triazine-resistant common ragweed has been identified. Initially, triazine resistance developed in fields where the triazine herbicides were used exclusively and where corn was grown as a monoculture without rotation. Weed populations that were originally mixtures of both susceptible and resistant biotypes gradually became the predominant, resistant types because the susceptible ones were controlled continuously. The infestation has spread as a result of feeding cows silage corn contaminated with the seeds of these resistant species and then spreading the manure back on the fields. Most fields associated with dairy production in New York State are probably infested with triazine-resistant common lambsquarters. Controlling the resistant species can be difficult, but the following suggestions may help:

1. Spread manure only on fields where an infestation already exists.

2. Rotate crops. This alters the cultural practices, changes the herbicides used, and employs different weed control strategies.
3. Rotate herbicides within crops. For example, when growing corn, select herbicide programs that do not rely entirely on triazines.
4. Cultivate to prevent weed escapes from forming seeds.

Herbicide Carryover and Soil Residues: Several vegetable herbicides can be excessively persistent even when used at rates suggested on the label. *AAtrex*, Sencor, Devrinol, and Treflan are examples of herbicides that often interfere with fall cover crops and sometimes injure susceptible vegetables the following spring.

To help avoid persistence, Cornell occasionally suggests dosages well below those stated on the label. Other ways of reducing danger from herbicide carryover are to apply two different herbicides at their lowest labeled rates; use low initial dosages and repeat if needed; use only extremely shallow incorporation when necessary; avoid applying persistent herbicides after late June or early July because degradation occurs mostly in the warmer months; and moldboard plow deeply before planting a succeeding crop.

Wildlife Damage Management

A vegetable grower can use a variety of nonchemical alternatives to reduce deer damage to crops. These techniques fall into three primary categories: exclusion, population reductions, and habitat modification. Fencing is the most common exclusion technique used to prevent deer damage. Woven-wire designs are the most effective physical barrier, with high-tensile, woven-wire fencing providing the ultimate in protection and durability. Deer can be successfully eliminated from large areas with an eight to ten foot high woven-wire fence. The advantages of this design are its effectiveness and low maintenance requirements after construction. Disadvantages include the high initial cost and the difficulty in repairing damaged sections.

A variety of multistrand, high-tensile, vertical, or sloped electric fence designs may effectively exclude wildlife. Electric high-tensile fences can be complete physical barriers, or more often, act as psychological deterrents. Deer can be excluded from crops with a five to six foot electric fence, even though they can easily jump over fences of this height. The most frequent reasons why electric fences fail to prevent deer damage include: the selection of an unsuitable fence design, the failure to install fencing according to specifications, and inadequate maintenance. High-tensile electric fences are more easily repaired than conventional fences, and may cost half as much as eight to ten foot woven-wire designs. Disadvantages include the need for frequent monitoring and vegetation control to maintain shocking power. Single-strand electric fences, combined with cloth strips treated with deer repellents or aluminium foil tabs coated with peanut butter to act as an attractant, and attached at three to four foot intervals along the fence, have successfully reduced summer deer damage to vegetables. High-visibility, electric polytape fences on fibreglass stakes provide another low-cost, portable design that can effectively reduce deer damage to vegetable crops.

Posting of private lands reduces the opportunity for sportsmen to harvest antlerless deer. Deer populations are regulated through the removal of breeding-age does. Growers who experience recurring deer damage should invite hunters to their property and mandate that they fill an antlerless tag (if available) before harvesting a buck. Reducing deer numbers on a unit-wide basis may lower crop losses. Deer depredation permits issued on a farm-by-farm basis have not controlled crop losses in other states.

Deer problems are most severe in fields near escape or resting cover. Mowing or removing brush in fields adjacent to crops may make the sites less attractive for deer and other problem wildlife. Some growers have experimented with lure crops to draw deer away from important harvestable fields, however, these efforts have had mixed success.

Repellents

Repellents may be cost-effective for controlling wildlife damage when light to moderate damage is evident, small acreage is damaged, and only a few applications will be needed for adequate control. If these three conditions are not satisfied, it is best to look at the cost-benefit ratios of alternative control measures.

With the use of repellents, some damage must be tolerated, even if browsing pressure is low. None of the existing repellents provide reliable protection when deer densities are high. Repellents should be applied before damage is likely to occur, when precipitation is not expected for 24 hours, and temperatures will remain between 40° to 80°F for that period. Hand-spray applications may be cost effective on small acreages, while machine sprays will reduce costs for larger areas. If the materials are compatible, spray costs may be reduced by adding repellents to pesticide sprays. If browsing pressure is severe, a long-term damage management program should be implemented. Such a program should include potential habitat modifications, reductions in animal numbers, and an evaluation of fencing alternatives.

Woodchucks

Woodchucks are game animals in New York and can be hunted throughout the year without limit. A hunting license is required to harvest woodchucks. Woodchucks causing damage can be destroyed without a license under New York Conservation Law. Consult your regional Department of Environmental Conservation (DEC) office if you have questions about a specific situation. Growers have usually relied on lethal controls to reduce woodchuck damage. Spring is the best time to use lethal controls, because adults are active and young animals may remain within their burrow at this time. In addition, burrows are more evident before annual vegetation conceals their entrances, and other wildlife are less likely to use burrows as shelter at this time.

Shooting or trapping methods can be used to remove problem woodchucks from fields containing edible crops. It may be illegal or unsafe to shoot woodchucks under some circumstances. Woodchucks can be captured using #2 leg-hold traps, #160 or #220 body grips, or live traps baited with apples and set near burrow entrances. Traps should be checked at least twice daily. Only live traps should be used where pets or livestock might be inadvertently captured.

Lethal controls have been reported to have limited success in controlling woodchuck populations. Twenty-eight percent of the burrows fumigated during one study were reoccupied within two days, and 83 percent of the burrows were reoccupied within two weeks after treatment. The rapid reoccupation of burrows may have resulted because the treatment was applied during the day when woodchucks are least likely to be inside burrows and because more than one animal may use a burrow system. Similarly, shooting and trapping methods will not eliminate woodchucks from a farm. In one Pennsylvania study, 1,040 woodchucks were removed from a 600 acre site over four years without significantly affecting the population's size.

Fencing can be effective at reducing woodchuck damage. Woodchucks may be excluded from small areas with a four-foot-high hardware cloth fence with the bottom buried ten to 12 inches, plus an electric hot-shot wire located four- to five- inches-high and an equal distance away from the outside of the fence. Electric, high-tensile deer fences may be modified to exclude woodchucks by adding additional wires at five to six inch intervals up to 18 inches high.

There are no products currently registered with the EPA for use as woodchuck repellents.

Modification of habitats in and around fields can help reduce their attractiveness to woodchucks, and can increase the effectiveness of other control techniques. Elimination of brush piles and overgrown areas will facilitate the discovery and treatment of burrows. Unfortunately, elimination of these areas may also eliminate valuable habitat for other wildlife. There is no research that indicates supplemental plantings can draw woodchucks away from more valuable crops.

Rabbits

Cottontail rabbits are found throughout New York State, excluding the central Adirondacks. Most rabbit damage occurs during spring or early summer when succulent new plants are available in crop fields or gardens.

Fencing can effectively protect small areas of high-value vegetable crops. Two-foot-high poultry-wire with 1- to 2-inch openings can adequately exclude rabbits. The fencing should be supported every six to eight feet with heavy wooden stakes or metal posts. It is critical to keep the lower edge of the fence tight to the ground, as rabbits will usually not dig under a fence, but will try to squeeze through existing openings or loose areas. If fences are installed in the spring and removed and stored in a dry place during the fall and winter, the materials may last five to ten years.

Cottontails are game animals in New York, and sport hunting can be used to effectively lower rabbit numbers in rural areas. Rabbits causing damage may be destroyed without a permit under existing New York State Environmental Conservation Law (Title 5, Section 11-0523). In more urban locations where hunting may not be possible, rabbits can easily be captured in box traps or commercial wire cage traps. Permits from DEC will be required to trap and transport live cottontails, and it is best to contact your Regional DEC office before capturing or killing rabbits. Live traps can be baited with apples or corn, and once a rabbit is caught in a trap, its lingering scent will often attract other cottontails. Putting a few cottontail droppings in a trap along with the bait will enhance the trap's effectiveness. Rabbits are usually most active just after sunset and just before sunrise. Live traps should be set prior to these peak activity periods.

One repellent is commercially available to protect carrots from rabbit damage. There are no research reports available to indicate the effectiveness of these materials for preventing rabbit feeding.

Cottontails require dense vegetation near feeding areas for protection from predators. Overgrown ditches, stream banks, fence rows, or brush piles within or near crop fields may harbour abundant rabbits. Mowing, cutting brush, and removing brush piles or overgrown areas may effectively control cottontail problems.

Raccoons

Raccoons like sweet corn when kernels are at the milk stage and will dig holes in watermelons to scoop out the contents. Foraging usually takes place at night.

Raccoons are protected furbearers in New York State, with established hunting and trapping seasons. All local laws or ordinances must be followed, and your DEC Regional office should be contacted before implementing lethal raccoon control.

Exclusion is the best method for preventing or reducing raccoon damage. Sweet corn or melon fields can be protected with double strand electric fences with wires five and ten inches above ground. Fences can be turned off during the day, or left on continually if woodchuck damage is also a problem. Fences should be installed at least two weeks before sweet corn reaches the milk stage, so the raccoons do not develop foraging habits.

No frightening methods have effectively reduced raccoon damage to sweet corn for more than short periods of time. Lights, radios, dogs, scarecrows, streamers, aluminium pans and lids, etc. have been tried with little success. Also, no toxicants or fumigants are currently registered for raccoon control. The repellents tested to date have also been ineffective.

Raccoons can be easily captured in traps. Live traps at least 10 by 12 by 32 inches in size and constructed of sturdy metal can be baited with marshmallows to reduce captures of nontarget species. Environmental Conservation Law permits only licensed Nuisance Wildlife Control persons or Wildlife Rehabilitators to transport live-captured animals, so the landowner must euthanize raccoons caught in cage traps or release them at the site of capture. Raccoons may also be captured in #1.5 leg-hold traps, and #160 or #220 body gripping traps can effectively catch raccoons in bait boxes. Body gripping traps are designed to kill quickly and should not be used in areas where there a chance of catching nontarget animals.

Due to the spread of raccoon rabies in New York, contact with raccoons should be minimised. Animals shot or captured in traps should only be handled with rubber gloves, and all traps and equipment should be disinfected with a bleach and water solution. Carcasses should be cremated or deeply buried. Field workers should be warned to avoid any wildlife which acts abnormally, as rabies may spill over into other mammal species.

Birds

General Information: Agricultural regions throughout North America are susceptible to bird damage. Bird species which impact vegetable growers fall into two broad categories: resident and migratory. Resident birds (i.e., crows, turkeys, etc.) may inflict low levels of damage throughout the grain-ripening period or pull up sprouting grain in the spring. Migratory species (i.e., starlings, blackbirds, etc.) often forage in large flocks and can cause severe corn losses in a few days.

All birds, except feral pigeons, house sparrows, and starlings, are protected by the Federal Migratory Bird Treaty Act. However, Section 21.43, Title 50 CFR, states that, "A Federal Permit shall not be required to control red-winged, rusty, and Brewer's blackbirds, cowbirds, all grackles, crows, and magpies when found committing or about to commit depredations upon ornamental or shade trees, agricultural crops, livestock, or wildlife, or when concentrated in such numbers and manner as to constitute a health hazard or other nuisance. . ." When considering harassment or lethal methods to control problem birds, local ordinances must still be followed, so it is best to consult with local law enforcement authorities or Regional DEC offices if you have questions concerning specific situations.

Exclusion: Netting is the most effective method for controlling bird damage. The cost of bird-proof plastic netting varies considerably with the type, manufacturer, and quality. The initial cost of netting may be quite high, but the cost can be prorated over the three to ten year life expectancy of the material. There is a high labour cost for installation and removal of netting, and this method is usually cost-effective only for the most valuable crops. Monofilament lines have been used in several spacings and configurations to reduce bird damage at landfills, fish hatcheries, public parks, and agricultural fields, although species responses are quite variable. For example, gulls, crows, and sparrows appear to be particularly sensitive to lines and have been successfully repelled. Usually lines are only practical for small plots or home gardens.

Auditory Frightening Devices

Broadcasts of recorded distress or alarm calls have been used successfully to drive birds from fields, orchards, and roosts. The sounds of a recorded propane explosion and distress calls were evaluated for reducing night heron damage to aquaculture ponds. The herons completely habituated to the cannon recording after six nights of exposure. However, more than 80 percent of the birds left the pond when the distress call was used, and no habituation occurred after six months. The use of recorded distress or alarm calls may have additional applications for bird management, but most calls are species specific, so growers must identify birds causing the damage.

Other bird-control methods (bangers, crackers, poppers, bombers, sirens, etc.) rely on perceived danger for their effect. These techniques have been effective for short-term control, as scare devices that produce sounds other than alarm/distress calls have no persistent effect on space use or food intake of birds. Best results are obtained when: sounds are presented at random intervals; a range of different sounds are used; the sound source is moved frequently; sounds are supported by other methods, such as distress calls or visual deterrents; and sounds are reinforced by real danger (i.e., shooting). Experiments conducted to date indicate that ultrasound probably cannot be heard by birds.

Visual Frightening Devices

Eye-spot balloons were reported to reduce damage to citrus groves. Balloons may be a more acceptable bird control practice in suburban areas where sound devices could lead to conflicts with neighbouring landowners; however, the effectiveness of balloons appears to depend on the bird species causing the damage, and results are variable.

Reflecting tapes placed three to seven yards apart were effective in protecting small fields (one acre) of sweet corn, millet, and sunflowers from bird damage. However, reflecting tapes spaced three to four yards apart did not protect blueberries from robin or starling damage. Reflecting tape requires substantial labour for installation and maintenance.

Cultural Practices

Fields near convenient perches, such as snags or power lines, may be prone to increased losses. Grass and weed control may reduce the numbers of certain seed-eating bird species. If possible, it is best to establish new vegetable fields away from cover or perch sites or to remove these attractive habitat features from areas next to existing plantings.

Insect control may reduce the number of birds feeding in corn fields. European corn borers attract red-winged blackbirds to sweet corn fields, and controlling corn borers will result in decreased bird damage to corn ears. Blackbirds generally feed on insects in corn fields during the silking period. Red-winged blackbirds accounted for more than 90 percent of bird observations in sweet corn fields in the Hudson Valley, and corn damage by red-wings did not occur until the ears reached the milk stage. Avian damage increased rapidly once kernels reached this stage, so advancing harvest by a few days can greatly reduce losses.

Cultivars of corn differed substantially in susceptibility to bird damage in free-choice trials. Husk coverage of the ear (expressed as husk weight or length of husk extension beyond the cob tip) accounted for about 68 percent of the variation between cultivars in aviary tests. In general, the most susceptible cultivars received about 2 1/2 times more damage than the least susceptible. However, stage of kernel maturity at the time of bird damage influenced the relative resistance of cultivars, as differences between cultivars were less pronounced at late stages of maturity.

Management Recommendations

The success of most bird damage management activities depends upon the interest and persistence of the field person conducting the control efforts. Field staff should monitor bird numbers and species present. If scare devices are warranted, they must be functional from shortly before sunrise until sunset. Birds quickly habituate to most frightening techniques, but the period of effectiveness can be lengthened by frequently moving scare devices, and reinforcing the stimuli with real danger (shooting), where legal and practical. Combining a number of different methods usually results in greater success than relying on a single technique.

Damage mitigation measures may only be cost-effective early in the harvest season when birds are concentrating their feeding on early-ripening cultivars. Growers should carefully monitor these fields, so timely and appropriate control efforts can be implemented. Control measures often become cost-effective when large flocks of migratory birds move into the area.

In particular, migratory blackbird flocks can inflict substantial damage during short time periods. Again, careful monitoring is necessary to determine the timing and potential amount of crop loss. Growers should decide whether or not to attempt to control damage, and which techniques are appropriate, based on their estimates of potential losses.

Chapter 7

Pesticide Information and Safety

Pesticide Classification and Certification

Putting it simply, a pesticide is substance used to control pests. Federal law, the Federal Insecticide, Fungicide and Rodenticide Act, or FIFRA, created two classifications of pesticides — general-use and restricted-use. General-use pesticides may be purchased and used by anyone, but only a certified applicator may purchase a restricted-use pesticide. Restricted-use pesticides must be used by a certified applicator or someone under their supervision.

The same federal law that classified pesticides divided applicators into two groups: private and commercial. Private Applicators use or supervise the use of pesticides to produce agricultural commodities or forest crops on land owned or operated by the Private Applicator or their employer.A farmer must be certified as a private applicator in order to purchase and use restricted-use pesticides on agricultural commodities. (No certification is needed if a farmer does not use restricted use pesticides.)

A Commercial Applicator uses or supervises the use of pesticides for any purpose or on any property not covered by the private applicator classification. In New York, a commercial applicator must be certified to purchase or use any pesticide whether it is general- or restricted-use.

Use Pesticides Safely

Using a pesticide imparts a great responsibility on the user to be a good steward of your health and that of others. Keep in mind that there is more to "pesticide use" than the application. Pesticide use also includes mixing; loading; transporting, storing or handling after the manufacturer's seal is broken; cleaning of pesticide application equipment; and any preparation of a container for disposal. All of these actions require thoughtful planning and preparation. They are also regulated by state and federal laws that are intended to protect the user, the community, and the environment from any adverse effects pesticides may cause.

Plan Ahead

Many safety precautions should be taken *before* you actually begin applying pesticides. Too many pesticide applicators are dangerously and unnecessarily exposed to pesticides while they are preparing to spray. Most pesticide accidents can be prevented with informed and careful practices. Always read the label on the pesticide container before you begin to use it. Make sure that you understand everything you need to know about the pesticide ahead of time so that you are a responsible user. Carefully follow all the directions and precautionary advice on the label. Be sure that you are prepared to deal with an emergency exposure or spill before you begin using pesticides. Be sure to know the first aid procedures for the pesticides you use.

Move Pesticides Safely

Carelessness in transporting pesticides can result in broken containers, spills, and contamination. Once pesticides are in your possession, you are responsible for safely transporting them. Accidents can occur, even when transporting materials a short distance. If a pesticide accident occurs, you are responsible. Do all you can to prevent a problem when transporting pesticides. Be prepared in case an emergency should arise.

Personal Protective Equipment

The need for personal protective equipment depends mainly on the pesticide being handled. Personal protective equipment requirements are printed on pesticide labels. These requirements are based on the toxicity, route of exposure, and formulation of that pesticide. The requirements posted on the label are the minimum that must be worn during the pesticide use. A pesticide user always has the option of wearing more protection than the label requires.

The activity, the environment, and the handler also influence the choice of protective equipment. The activity-related factors are type of activity, duration of the activity, equipment, and deposition pattern of the pesticide onto the handler. Mixing/loading procedures often require extra precautions when the pesticide is in concentrated form. Studies show that you are at a greater risk of accidental poisoning when handling pesticide concentrates. Pouring pesticide concentrates from one container to another is the most hazardous activity. A closed mixing/loading system can reduce this risk. Closed mixing systems are part of an array of protective devices call "engineering controls."

Avoid Drift, Runoff, and Spills

Pesticides that fall anywhere but on the target area can injure people, crops, and the environment. Choose weather conditions, pesticides, application equipment, pressure, droplet size, formulations, and adjuvants that minimise drift and runoff hazard.

Avoid Equipment Accidents

Properly maintained and carefully used equipment contribute to safe pesticide application:

- Be sure to turn off your machinery before making any adjustments.
- Do not allow children, pets, or unauthorised people near the pesticide equipment.
- Between jobs, depressurize tanks or systems.
- Always return equipment to appropriate areas for cleaning and storage when pesticide applications are completed.

Pesticide Storage

Most pesticide applicators use existing buildings or areas within existing buildings for pesticide storage. Whether you choose a site to build a new storage area or use existing buildings, you need to consider several points:

- The site should be in an area where flooding is unlikely.
- It should be downwind and downhill from sensitive areas such as houses, ponds, and play areas.
- There should be no chance that runoff or drainage from the site could contaminate surface or groundwater.

Storage facility check list:

- Is the facility separated from offices, workshops, and livestock areas?
- Is the facility separated from wells, streams, lakes, ponds, wildlife?
- Is the facility separated from food and feed?
- Is the facility made of fire resistant building materials?

- Does the facility have impermeable flooring?
- Does the facility have liquid spill containment (berms to hold 25% of liquid storage)?
- Can the doors be locked?
- Is the facility fenced in?
- Are warning signs posted?
- Is a spill kit readily available?
- Are fire extinguishers readily available?
- Is personal protective equipment readily available?

Protect Honey Bees from Insecticides

Honey bees, wild bees, and other insects are important for proper pollination of many crops. Vine crops, for example, must be pollinated because they have male and female flowers, and pollen must be transferred from the male to female flowers if fruit is to set. Poor pollination results in small or odd-shaped fruit as well as low yields. Each flower must be visited eight or more times for adequate pollination to occur. To avoid harming bees with insecticide treatments, remember these points:

- Do not spray crops in bloom;
- Mow blooming weeds before treatment or spray when the blossoms are closed;
- Avoid application during the time of day when field bees are most numerous;
- Make application in the early morning or evening; and
- Always read the label before use.

If pesticides that are highly toxic to bees are used in strict accordance with label directions, little or no harm should be done to bees. Label statements on pesticides that are highly toxic to honey bees may carry a caution statement such as: "This product is highly toxic to bees exposed to direct treatment or residues on blooming crops or weeds. Do not apply this product or allow it to drift to blooming crops or weeds if bees are visiting the treatment area."

New York State Pesticide Control Legislation

Restricted Pesticides

In accordance with New York State law, pesticides that are highly toxic or that are persistent and accumulative are placed on a restricted-use list. These pesticides may be sold to and used only by certified applicators or used under the direct supervision of a certified applicator.

"Restricted-use pesticide" or "restricted pesticide" means any pesticide: listed in Section 326.2 of Part 326 of the NYSDEC regulations, that meets the criteria of Environmental Conservation Law Section 33-0101(42), or whose labelling bears the statement "Restricted-Use Pesticide."

The NYSDEC uses the following criteria and designations (A through H) to classify those pesticides restricted for use, sale, purchase, or distribution in New York State as noted within Part 326.2:

- May be distributed, sold, purchased, possessed, and used only upon issuance of a commercial or purchase permit for any uses listed on the approved label as registered with the NYSDEC.
- May be distributed, sold, purchased, possessed, or used only upon issuance of a commercial permit or purchase permit for those purposes listed in "Part 326 – Registration and Classification of Pesticides," a publication of the NYSDEC.
- The commissioner of the NYSDEC may place any conditions on the registration of any product that are deemed necessary to prevent damage to health, property and wildlife. Classification as 'restricted-use' may be necessary.
- Federally restricted pesticides per EPA.
- Any product whose label limits use to commercial pesticide applicators only may be distributed, sold, purchased, possessed, and used only upon issuance of a commercial permit or certification identification card. Label statements that limit use to commercial pesticide applicators include but are not limited to the following: (1) only for sale to and use and storage by commercial pest control operators and (2) to be applied only by or under the direct supervision of commercial applicators responsible for insect control program.
- Any pesticide labeled for direct application to or in surface waters may be distributed, offered for sale, sold, purchased, possessed, or used only by the holder of a valid commercial permit, certification identification card, or purchase permit.

Federally restricted pesticides will state "RESTRICTED-USE" on the label. Pesticides restricted by the NYSDEC may not include such a label statement. Always check the restricted-use status of the pesticide before purchasing or using it to determine any restrictions placed on the product.

Note: Use of federally restricted pesticides by an individual under the instruction and control of a certified private pesticide applicator requires on-site direct supervision by a certified applicator according to NYS regulation (6 NYCRR part 325.7(d)(2)(ii)). Certified commercial pesticide technicians are not permitted to purchase restricted-use pesticides.

The NYSDEC may also determine that a pesticide can not be used in a county or other area of New York State where it poses an unacceptable risk. For example, certain pesticides are prohibited from use on Long Island due to groundwater or other concerns. Pesticides with additional use restrictions will have a label statement or statements indicating which counties or areas the pesticide may not be used in.

Pesticide Product Registration Information for New York State

Cornell University and the NYSDEC provide access to several databases that allow for retrieval of pesticide information, including product registrations and pesticide label images pertinent to New York State.

Verifying Pesticide Registration and Restricted-Use Status

Any pesticide used in New York State must be currently registered with the NYSDEC and the USEPA. Applicators can easily verify whether pesticides are currently registered and classified as restricted-use in New York State if they have access to the Internet. Although some pesticide registrations at this web site may be shown as expired, we are advised by the NYSDEC that if a pesticide is listed as currently registered, it is considered approved for use in New York as of the list date.

Pesticide Recordkeeping/Reporting

New York State Requirements

Manufacturers and importers. The NYSDEC requires annual reports from manufacturers and importers documenting all restricted-use product sales within the state. These records include the EPA registration number, container size, and number of containers sold to New York purchasers. Records need to be maintained for at least three years.

Commercial applicators. The NYSDEC requires annual reports from commercial applicators and technicians about their pesticide use. (Both restricted- and general-use pesticides are included in the report.) Annual reports include the EPA registration number, product name, quantity of pesticide used, date applied, and location of application by address (including 5-digit zip code). All commercial applicators and technicians must file annual reports, even if no pesticide applications were made during the calendar year.

Commercial applicators and technicians must also maintain corresponding application records of the dosage rates, methods of application, and target organisms. These records need to be maintained annually, retained for at least three years, and be made available for inspection on request by the NYSDEC.

Reporting at point of sale. The NYSDEC requires every person who sells or offers for sale restricted-use pesticides to private applicators to report annually each sale of a restricted-use pesticide or general-use pesticide used in agricultural crop production to such applicator. This report includes: EPA registration number; product name of pesticide purchased; quantity purchased; date purchased; and location of intended application by address (including 5 digit zip code) or if address is unavailable, by town or city (including 5 digit zip code). An annual report is required even if no pesticides were sold during the calendar year.

Recordkeeping by private applicators. New York State regulations require private applicators to maintain restricted-use pesticide records. These records need to include, at a minimum, the restricted-use pesticide(s) purchased; crop treated; application method, and date(s) of application(s). These records need to be retained for at least three years and must be made available for inspection by the NYSDEC upon request.

Federal Private Applicator Recordkeeping Requirements

Federal regulations require private applicators to keep records of the restricted-use pesticides they apply. Federally required records need to contain the following elements:

- Applicator name;
- Applicator certification number;
- Month, day, and year of application;
- EPA registration number for product applied;
- Brand or product name of pesticide applied;

- Crop, commodity, or site treated;
- Size of area treated;
- Total amount of pesticide product applied; and
- Location (not the address) of the application using any of the following:
  - o County, range, township, and section;
  - o Maps or written descriptions;
  - o A USDA identification system, which involves maps and a numbering system to identify field locations; or
  - o The legal property description.

Under the federal recordkeeping regulations, spot applications are applications of a restricted-use pesticide to less than 1/10 of an acre on the same day. Spot application records need to include: month/day/year of application; brand or product name of pesticide applied; EPA registration number; total amount of pesticide applied; and location of the application designated as a "spot treatment" with a brief description. NOTE: Spot treatments do not apply to nurseries or greenhouses. Applications at these locations need to be recorded as if they are a non-spot application.

Federally required records must be made within 14 days of the application and need to be maintained for 2 years following the application. Access to the record information is limited to: USDA-authorised representatives presenting identification; state-authorised representatives presenting identification; and attending licensed health care professionals (or those acting under their direction) when treating individuals who may have been exposed to a restricted-use pesticide.

EPA Worker Protection Standard (WPS) for Agricultural Pesticides

The Worker Protection Standard contains requirements designed to reduce the risks of illness or injury resulting from occupational exposures to pesticide handlers and agricultural workers. Accidental exposure of agricultural workers and other persons to pesticides used in the production of agricultural plants on farms, nurseries, greenhouses, and forests are included under these regulations. WPS requirements include the following:

- Restricted-entry intervals (REIs) for most pesticides.
- Personal protective equipment (PPE) for handlers and early-entry workers.

- Decontamination supplies and emergency assistance.
- Pesticide safety training and posting.
- Revised labelling that includes specific WPS instructions.

Reduced-Risk Pesticides, Minimum-Risk Pesticides, and Biopesticides

Reduced-risk Pesticides

Since 1993, EPA has expedited the registration of conventional pesticides with characteristics such as very low toxicity to humans and nontarget organisms including fish and birds, low risk of groundwater contamination or runoff, low potential for pesticide resistance, demonstrated efficacy, and compatibility with IPM. Materials meeting these criteria are referred to by EPA as "reduced-risk." The "reduced-risk" designation applies only to certain uses of a particular pesticide and may not include all label uses for that product.

Minimum-risk Pesticides

Minimum-risk pesticides are certain products that are exempted from EPA registration and therefore have no EPA registration number. They contain only active ingredients outlined in FIFRA 40 CFR 152.25(g) ("the 25b list") and inert ingredients currently identified in Federal Register Notice 59 FR 49400 (List 4a).

New York State pesticide registration policy exempts true minimum-risk pesticides from product registration requirements. Policies may differ in other states; some states require such products to carry a state registration number.

Biopesticides

Biopesticides, or biological pesticides as defined by EPA, are certain types of pesticides derived from natural materials such as animals, plants, bacteria, and certain minerals. These include microbial pesticides (which contain bacteria, fungi, virus, etc., as the active ingredient); plant-pesticides (pesticidal substances that plants produce from added genetic material such as corn genetically modified to produce *Bacillus thuringiensis* toxins); and biochemical pesticides comprised of naturally occurring substances that control pests by nontoxic mechanisms (such as pheromones or some insect growth regulators). Biopesticides must be registered with EPA.

FIFRA 2(ee) Recommendations

Certain limited variations from the use directions specified on pesticide labels are authorised under FIFRA Section 2(ee).

These "2(ee) recommendations" allow:

1. Use at any dosage, concentration, or frequency less than specified on the labelling.
2. Use against any target pest not specified on the labelling.
3. Methods of application not prohibited on the labelling.
4. Mixtures with fertilizer unless prohibited on the labelling.

In New York State, all 2(ee) recommendations must be approved in writing by the NYSDEC with the exception of those pesticides applied for agricultural purposes at less than label rate and mixtures with fertilizer (item 4 above).

No fee is required for a 2(ee) recommendation request.

Requirements:

1. Anyone receiving approval of a 2(ee) recommendation is responsible for distributing the recommendation to all users of the product pursuant to the approved recommendation.
2. Any user must have the 2(ee) recommendation in his or her possession at the time of application, e.g., "Cornell Guidelines."

Information required for submitting 2(ee) recommendations:

1. Requests for approval of 2(ee) recommendations must be made in writing by recognised research institutions, certified crop advisers, manufacturers (registrants), or organisations representing individual users. Individual users may not request approval of 2(ee) recommendations.
2. Requests must be accompanied by data demonstrating that the use will be effective.

Protecting Our Environment

Environmental Impact Quotient

Because of the pesticide registration process, there is a wealth of toxicological and environmental impact data for most pesticides that are commonly used in agriculture. However, these data are not all readily available or organised in a manner that is usable to the IPM practitioner. A method called the Environmental Impact Quotient

(EIQ) has been devised to calculate the environmental impact of most common pesticides (insecticides, fungicides, and herbicides) used in agriculture. The values obtained from these calculations can be used to compare the environmental impact of different pesticides and pest management programs.

The EIQ is based on data obtained from EXTOXNET, Cornell Pesticide Management and Education Program, SELCTV, the National Pesticide/Soils Database developed by the USDA Agricultural Research Service and Soil Conservation Service, the 1989 New York State Pesticide Recommendations, Material Safety Data Sheets (MSDS), and technical bulletins developed by the agricultural chemical industry.

Factors such as toxicity (dermal, bird, chronic, bee, fish, beneficial arthropod), soil half-life, systemicity, leaching potential, plant surface half-life, surface loss potential, and farm worker, consumer, and ecological effects are all considered when calculating an EIQ for a particular pesticide. The result is a single number describing the EIQ of a pesticide active ingredient. The New York State Food and Life Sciences Bulletin No. 139 entitled "A Method to Measure the Environmental Impact of Pesticides" describes in detail the derivation of the EIQ.

Once an EIQ value has been established for the active ingredient of a pesticide, field use ratings can be calculated. To compare pesticides and pest management strategies, the dose, the formulation or percent active ingredient of the product, and the frequency of application of each pesticide needs to be determined. To account for different formulations of the same active ingredient and different use patterns the EIQ Field Use Rating was developed. This rating is calculated by multiplying the EIQ value for the specific chemical by the percent active ingredient in the formulation and by the rate used (usually in pints or pounds of formulated product per acre).

EIQ Field Use Rating = EIQ x % active ingredient x rate

The lower the EIQ Field Use Rating the lower the environmental impact. This method allows comparisons of the environmental impact among pesticides and different pest management programs. For example, if several pesticides are registered against a particular pest, which pesticide has the least impact on the environment?

Table: Hypothetical example of the EIQ Field Use Rating for three different insecticides

Material	*EIQ*	*AI*	*Rate*	*EIQ field use rating*
*endosulfan**Thiodan 50WP	42.1	0.50	3.0	63.15
*carbaryl*Sevin 50WP	21.7	0.50	6.0	65.1
*azinphos methyl*Guthion 35WP	44.9	0.35	2.2	34.6

Note: These products are no longer registered for use in New York State. They are shown here for illustrative purposes only.

Although *carbaryl* has a lower EIQ than *endosulfan* or *azinphos-methyl* it may take more applications to provide equivalent control. For example, 6 lbs/acre of Sevin may provide the same level of control of a certain pest as 3 lbs/acre of Thiodan or 2.2 lbs/acre of Guthion. In this situation, Guthion would have the lowest EIQ Use Rating and would have the lowest environmental impact. The EIQ Field Use Rating can be used to compare different pest management strategies. To compare different pest management programs, EIQ Field Use Ratings and number of applications throughout the season are determined for each pesticide and these values are then summed to determine the total seasonal environmental impact of the particular strategy. By using the EIQ model, it becomes possible for IPM practitioners to rapidly and easily estimate the environmental impact of different pesticides before they are applied.

Reducing Risk of Pesticide Exposure through use of Engineeering Controls

Why use Engineering Controls?

Because handling and applying pesticides is risky business, keeping pesticide exposure to a minimum should be a chief concern of any pesticide applicator. To reduce the risks associated with handling and applying pesticides, devices known as engineering controls can be used that help to reduce or practically eliminate exposure to toxic chemicals.

Loading the Sprayer

Closed Transfer Systems: Closed transfer systems allow concentrated pesticide to be moved from the original shipping container to the sprayer mix tank with minimal or no applicator contact. Many systems provide a method to measure the concentrated pesticide. Some systems also include a container rinsing system. Currently available closed transfer systems use a probe inserted into the pesticide container, a connector on the container that mates to a similar connector on the application equipment, or a vacuum-type (venturi) system that uses flowing water to transfer the chemical from the container.

Induction Bowls: Induction bowls are metal, plastic or fibreglass hoppers attached to the side of the sprayer or the nurse tank that allow pesticides to be added to the mix tank without the applicator climbing onto the spray rig. Pesticides are poured into the bowl and water is added to flush out the bowl and carry the pesticide to the spray tank. Often a rinse nozzle is mounted inside the bowl for rinsing out empty pesticide containers. Typically induction bowls are raised out of the way during spraying and lowered to about 3 feet above ground when loading the sprayer.

Direct Pesticide Injection System: Direct pesticide injection systems allow pesticides to be mixed directly with water in the sprayer plumbing system rather than in the main spray tank. The pesticide is pumped from its container and mixed with the water either in a manifold or at the main water pump. Only clean water is held in the main tank of the sprayer. An electronic controller and up to four pumps adjust the amount of concentrated pesticide that is injected into the water stream, allowing for variable application rates.

Container Rinse System: Container rinse systems consist of a rinse nozzle and a catch bowl that traps the container washings (rinsate). The empty container is placed over the rinse nozzle and a jet of water cleans the inside of the container. The rinsate caught in the bowl is pumped into the spray tank to be used along with the spray mixture. Often rinse nozzles are installed in chemical induction bowls.

Reducing Contamination at the Boom

Boom Folding/Extending: Manually folding booms can be a major source of operator contamination because the boom can be covered with pesticide from drift or dripping nozzles. Consider the use of hydraulic or mechanical folding methods.

Diaphragm Check Valves: Typically, when a sprayer is shut off and as the system pressure drops, any liquid remaining in the boom piping drips from the nozzles, possibly dripping onto the boom or even the operator. Diaphragm check valves installed at each nozzle prevent this by using a spring-loaded rubber diaphragm to close off the flow of liquid once the system pressure drops below about 10 pounds per square inch. When the sprayer is switched on and system pressure builds up, the valve opens and allows the liquid to flow through the nozzles.

Multiple Nozzle Bodies: Contamination can occur when operators change or unclog nozzles during an application. Multiple nozzle bodies (or turret nozzles) allow operators to switch between nozzles with a turn of the nozzle body rather than having to unscrew or undo a threaded or a bayonet fitting.

Hand Wash Water Supply - Providing adequate wash water is essential (and often required). A simple container with a hand-operated valve can be mounted on the side of the sprayer to provide clean water for hand washing and personal hygiene.

Drift and Contaminated Clothing in Cabs

Cab Filtration Using Carbon Filters: Carbon filtration systems are used to remove pesticide odour and pesticide-laden mistfrom fresh air used in a tractor or self-propelled sprayer cab. Carbon filtration systems are often a standard feature on self-propelled sprayers. Now many factory installed tractor cabs offer optional filtration systems. In 1998, the American Society of Agricultural Engineers (ASAE) adopted testing standards for operator cabs used in pesticide application. Cabs certified under this standard meet the requirements for enclosed cabs contained in the Worker Protection Standard.

Protective Clothing Lockers: To prevent contamination of the tractor or sprayer cab interior, protective clothing should be removed before entering the cab. A few sprayer companies offer a simple compartment (or locker) mounted to the side or front of the sprayer where protective clothing can be stored. Alternatively a locker can be fitted to the mixer wagon.

Controlling Drift

Low-Drift Nozzles: Low-drift nozzles create larger-size droplets than conventional nozzles. The larger droplet sizes are less prone to drift, reducing environmental and operator contamination.

Air Induction (Twin Fluid) Nozzles: These nozzles allow air to mix with the spray liquid, creating large, air-filled droplets that have virtually no fine, drift-prone droplets. The droplets explode when they contact their target and offer similar coverage to droplets from conventional, finer-spray nozzles.

Cleaning the Sprayer

Tank Rinse Systems: Tank rinse systems consist of a clean water supply tank mounted to the sprayer and one or more rotating discs or nozzles mounted inside the main sprayer tank. Water is pumped from the clean water tank to the rinse nozzles, which spray water around the inside of the spray tank. These systems are designed for infield rinsing of the sprayer so that the tank washings can be applied to the field at label rates.

Preparing the Crop Sprayer for Work

Checking the Sprayer

Surveys have shown that many farmers are using inaccurate sprayers; faulty sprayers contribute to increased drift levels and waste money through inefficiency and overuse of chemicals. Sprayers must be regularly checked to ensure that proper maintenance has been carried out and that no outstanding repairs need to be done. Before attempting any work on a machine make sure that it is fully supported on stands and that all necessary protective clothing is on hand.

The cost of replacing a faulty gauge which has been indicating at 15% below the actual pressure is recouped in around two hours' operation.

Maintenance measures such as fitting a new set of nozzles at the beginning of each season also save money. Even when there is overdosing by as little as 5%, the cost of a new set of nozzles would be recovered in less than a day's work.

Caution

- Take great care when adjusting a sprayer while the tractor engine is running
- Engage the handbrake when leaving tractor seat.

Fitting the Sprayer to the Tractor

The selected tractor must always be powerful enough to operate the sprayer efficiently under the working conditions that will be encountered. All its external services - hydraulic, electrical and pneumatic - must be clean and in working order. Tractors fitted with cabs must have efficient air filtration systems. All protective guards must be in place. Trailed sprayers are often close-coupled to the tractor, so it is essential that the drawbar and the PTO shaft are correctly adjusted for turning. PTO shafts must be disengaged when making very tight turns.

Checking the Operation of the Sprayer

Part fill the tank with clean water and move the sprayer to uncropped waste ground. Remove the nozzles. Although not using any chemical at this point, get into the habit of wearing a coverall, gloves and a face visor when working with the sprayer. Engage the PTO and gently turn the shaft, increasing speed slowly to operating revs. Test the on/off and pressure relief valves, and check the agitation system. Flush through the spray lines, then switch off the tractor. Refit the nozzles and check the liquid system again for leaks.

Pre-season Maintenance

- Follow the checklists before you begin spraying

Hoses

Have you checked:

- for splits and cracks, particularly where booms fold
- connections to ensure they are water-tight
- for hose chafe, particularly in routing clips

Action: Replace damaged hoses.

Filters

Have you checked:

- for missing filter elements and seals
- for leakage
- for blocked or damaged filters

Action: Replace any damaged or blocked filters.

Tank

Have you checked:

- for fractures and any other damage
- the tank sits firmly in its mount
- the securing straps are correctly adjusted
- the agitation is working
- the tank is clean .

Controls

Have you checked:

- the control circuitry (electrical, hydraulic or air) for correct operation
- valves for both internal and external leaks

Action: Replace leaky valves, which waste money and are potentially dangerous to operators and the environment.

Pump

Have you checked:

- lubrication levels
- for leakages
- the air pressure in the pulsation chamber (if fitted) is at the recommended level
- the pump rotates freely without friction or noise. Do so by rotating manually or starting at low speed (corrosion may cause seizing up)

Pressure Gauge

The pressure gauge is vital for indicating whether the nozzles are delivering the correct amount of chemical per unit time while spraying.

If you have any doubts about the pressure gauge, replace it or refer the problem to the manufacturer or supplier.

Boom

The boom suspension must operate smoothly to ensure a stable 'nozzle platform'.

Ha*ve you checked*:

- friction plates, dampers, shock absorbers, pivots and slides - and scheduled future maintenance
- boom movement and stability*
- the boom folding mechanism
- the height adjustment mechanism for correct functioning
- for distortion or sagging
- the break backs for correct operation.

**The boom movement must be free but not too loose: for a 40 foot boom, the tip should return to within 2" of its original height after a vertical displacement of 12" or more.*

Boom Piping

Have you checked:

- the condition of all pipework
- the nozzle bodies for damage or loose fit
- for any damaged units, and replaced them
- for leaks under pressure .

Action: remove boom end caps and flush clean water through.

Anti-drip valves

Avoid leaks by keeping these small valves, above or to one side of each nozzle, in good repair. Damaged diaphragms and seats will lead to faulty functioning.

Examine all units under pressure to test for leaks and correct cut-off action.

Nozzles

Have you checked:

- all nozzles on the boom are the same
- all nozzles are in good condition, with no evidence of streaks or irregularities in the spray pattern
- all nozzles are clean and free from obstruction (note: clean with a soft brush or airline - don't damage nozzles by using wires or pins)
- all nozzles deliver to within + or - 5% of the manufacturer's chart value

Action: Always ensure the correct nozzles and operating pressure are selected before use. Have two or three sets of nozzles in stock to meet different spray qualities at different volume rates. Inspect nozzles throughout the season to avoid faults which could prove both costly and damaging to the environment if they develop unchecked. Using water only, set to 'spray' at the specified pressure and collect the output from each nozzle in turn for a period of 60 seconds. Record each output and replace those outside the 5% tolerance around the manufacturer's chart value.

Calibration

Equipment must be calibrated accurately if correct dosages are to be applied. The sprayer must give the right total gallonage per acre, and each nozzle must deliver its share of that total. The pesticide may be ineffective or may be wasted if it is not applied as accurately as possible. Over application results in more costly operation or causes off-target environmental contamination. Under application of materials may result in poor control and could require repeated applications which will be more costly.

The United States Department of Agriculture and the Environmental Protection Agency recommend that pesticides be applied within ten percent of the intended rate of application. Frequent checks of sprayer and tractor performance should be made to ensure that the machines are operating as expected. As the recommended label rates of active ingredient required per acre decrease, it is even more important that operators give tractor and sprayer performance proper attention. There are many different ways to calibrate liquid sprayers. Some growers actually spray a measured acre; others spray a small part of an acre; others keep the sprayer stationary, but collect and measure the spray for a certain number of minutes. If the initial results show that the volume of water is too low, the gallonage can be increased by reducing tractor speed or changing to larger nozzle tips. If the gallonage is too high, the reverse can be done. It is not desirable to make drastic changes in pressure in order to adjust gallonage per acre. Most nozzles have a relatively narrow pressure range in which they will operate properly.

Only clean water should be used when calibrating sprayers. Calibration requires knowledge of the amount of material applied over a known area. Liquid chemical application requires information on the volume (gallons) of material applied. The equipment required to calibrate a broadcast sprayer using this technique includes a tape measure, a stopwatch, and a measuring jar graduated in ounces. Personal safety protection equipment should also be worn when working around the sprayer. The most accurate calibrations are performed by measuring the output of each nozzle (gallons per minute, gpm), measuring nozzle pressure (pounds per square inch, psi) with a pressure gauge on the spray boom, and measuring travel speed (miles per hour, mph) in the field with the sprayer operating.

Travel Speed Calibration

Travel speed is a critical factor in maintaining accurate application rates and will influence spray deposition depending on location within the canopy. The slower a sprayer travels, the greater the uniformity in spray deposition. Although there is inconsistency in research results that try to determine the effect of travel speed on average spray deposition, all studies to date have been in agreement that the higher the travel speed, the greater the variability in spray deposit. Variation in spray deposit is an important factor where uniformity of spray coverage throughout the canopy is required. Conclusions from research were drawn using travel speeds of 1-4 mph.

Factors that will affect travel speed include:

- weight of sprayer to be pulled
- slope of terrain
- ground conditions travelled over (wheel slippage!)

The best way to measure travel speed is to pull a sprayer with tank half filled with water on the same type of terrain that the sprayer will be operated on. Set up test course at least 100 feet long, measure the course with a tape measure. Do not pace the distance. The longer the course the smaller the margin of error. Run the course in both directions.

Use an accurate stop watch to check the time required to travel the course in each direction. Average the two runs and use the following formula to calculate the speed in MPH.

Formula:

Your figures:

Tractor gear__________ Engine revs.__________

$$\text{MPH} = \frac{\text{ft traveled}}{\text{sec. traveled}} \times \frac{60}{88} = \frac{\text{sec.}}{\text{traveled}}$$

Boom Sprayer Calibration

Step 1. Record your tractor sprayer speed from above:

MPH = ________

Step 2. Record the inputs:

	Your Figures	*Example*
Nozzle type on your sprayer (all nozzles must be identical)	________	110 04 flat fan
GPA (from manufacturer's label)	________	20 GPA
Measured sprayer speed	________	4 MPH
Nozzle spacing	________	20 inches

Step 3. Calculate the required nozzle output:

$$\text{Formula : GPM} = \frac{\text{GPA x mph x nozzle spacing}}{5940\ (\text{constant})}$$

$$\text{Formula : GPM} = \frac{20 \text{ x } 4 \text{ x } 20}{5940} = \frac{1600}{5940} = 0.27\text{GPM}$$

$$\text{Your figures : GPM } = \frac{\text{X} \qquad \text{X}}{5940} = \frac{\quad}{5940} = \_\_\_\_ \text{ GPM}$$

$$\text{Formula : GPM} = \frac{\text{GPA x mph x nozzle spacing}}{5940\ (\text{constant})}$$

Step 4. Operate the sprayer:

- Set the correct pressue at the gauge using the pressure regulating valve

- Collect and measure the output of each nozzle for one minute
- The output of each nozzle should be approximately the same as calculated in Step 3 above
- Replace all nozzle tips which are more than 10% inaccurate.

Banded Boom Sprayer Calibration

The only difference between the boom sprayer calibration mentioned above and calibrating for a banded sprayer is the input value used in the formula in Step 3.

For a single nozzle banding applications:

Nozzle spacing = sprayed band width or swath width (in inches)

For multiple nozzle directed applications:

Nozzle spacing = row spacing (in inches) divided by the number of nozzles per row.

Minimally, vegetable sprayers should be calibrated at the beginning of the spraying season. Accuracy is important to any calibration technique. The more accurate the measurements, the more accurate and reliable the calibration calculations. If nozzle output (gpm) at the desired operating pressure varies more than ten percent from the manufacturers specification, the nozzle should be replaced. Volume measurements should be made to the nearest 1/2 ounce when collecting small volumes of water. Time should be measured to the nearest 1/10 second. Nozzle performance should be checked frequently because nozzles may wear, especially when operated at high pressures or if used to apply abrasive materials. Local Cooperative Extension offices can supply additional information about the details of sprayer calibration.

Minimising Pesticide Drift

Spray drift of pesticides is an important and costly problem facing pesticide applicators. Drift results in damage to susceptible off target crops, environmental contamination to watercourses and a lower than intended rate to the target crop, thus reducing the effectiveness of the pesticide. Pesticide drift also affects neighbouring properties, often leading to concern and debate.

There are two types of drift, airborne drift, often very noticeable and vapour drift. The amount of vapour drift will depend upon atmospheric conditions such as humidity, temperature and the product being applied and can occur days after an application is made. Drift is influenced by many interrelated factors including droplet size, nozzle type and size, sprayer design, weather conditions and last but not least the operator.

Droplet Size

In the past crops were drenched with high volumes and coarse droplets at up to 50 gallons per acre resulting in plants dripping with excess pesticide. The belief that too much is better than too little is misplaced. Dripping crops lead to environmental pollution such as soil contamination and an excessive number of tank loads per acre results in poor time management.

Lower volumes must be used which may result in smaller droplets although there is a limit to droplet size because of concerns about drift. Droplets under 150 microns generally pose the greatest hazard; droplets less than 50 microns have insufficient momentum for impaction as they remain suspended in the air indefinitely or until they evapourate. Research in England concluded that a 100 micron droplet takes about 11 seconds to fall ten feet in still air; when a similar size droplet is released into a 5mph wind it will drift about 75 feet before hitting the ground. The higher the operating pressure, the smaller the droplet, conversely, low pressure produces large droplets that may bounce off the target. Certain spray surfactants can change the droplet spectrum, reducing the number of driftable droplets.

Nozzle Type and Size

Correct nozzle selection is one of the most important yet inexpensive aspects of pesticide application. A nozzle's droplet size spectrum determines deposition and drift. Conventional flat fan nozzles fitted to a crop sprayer produces droplets in the range of 10 - 450 microns. There are 25,000 microns in one inch. Drift is a major problem with droplets less than 100 microns.

Increasing the Volume Median Diametre (VMD) will certainly reduce drift, but too large a droplet will bounce off the leaves to the ground, thus causing pollution, wasting money and resulting in less product on the target. Drift has been a major concern for some years, off target application wastes money, reduces deposition on the target plant, pollutes water courses and may cause nausea to other people.

Conventional Flat Fan Nozzles

Nozzles with 80^0 degree angle produce coarser droplets than 110^0 at the same flow rate but 80^0 nozzles require the boom to be set at 17-19 inches whereas 110^0 nozzles can be set lower at 15-18 inches above the target. (The lower the boom, the less chance of drift). Spray quality is fine - medium at 15- 60 psi

Pre-orifice Flat Fan Nozzles

The internal design of this nozzle reduces the internal operating pressure compared to a conventional flat fan resulting in coarser droplets (high pressure creates fine droplets, low pressure creates coarser droplets). Available as 80^0 or 110^0 nozzles. Spray quality is medium - coarse at 30- 60 psi. Drift-guard is a well-known trade name.

Turbo-teejet

A turbulence chamber produces a wide angle flat spray pattern of 150^0. Spray quality is medium - coarse at 15- 90 psi. Nozzles can be set at 15-18 inches above the target.

Air Induction Nozzles

Air induction, air inclusion or venturi nozzles are flat fan nozzles where an internal venturi creates negative pressure inside the nozzle body. Air is drawn into the nozzle through two holes in the nozzle side, mixing with the spray liquid. The emitted spray contains large droplets filled with air bubbles (similar to a candy malt ball) and virtually no fine, drift-prone droplets. The droplets explode on impact with leaves and produce similar coverage to conventional, finer sprays.

Air induction nozzles reduce drift even at higher pressures of 80-90 psi. They are only available at 110 fan angles so boom height may need to be adjusted to 15-18 inches. The use of adjuvants will certainly help create bubbles and trials in Europe confirm this.

Air induction nozzles are regarded as expensive, the list price is three times the cost of a conventional flat fan tip, although some nozzle suppliers in NY will supply for approximately twice the cost.

Manufacturers Include

Greenleaf TurboDrop nozzles consist of two primary components - the venturi air aspirator and the exit pattern tip. A ceramic orifice in the venturi determines the flow rate of the complete assembly. The venturi is ISO colour coded to designate flow rate. The exit pattern tip does not affect flow rate: it is only used to form the desired spray pattern. Pressure range is 40-90 psi.

Spraying Systems Tee Jet Air Induction (AI) comprises a plastic body with a steel tip, rated for 30 to 100 psi. They are plastic, single-piece construction.

Hardi air induction is similar in construction to Spraying Systems AI nozzle, it is a one piece plastic nozzle.

Table 1: Comparison of nozzle type and air-assistance in reducing drift

Year	*Crop-free Buffer zone (ft)*	*Year of tests*	*Nozzle type ##*	*Boom height (inches)*	*Air Assistance*	*Drift Deposition (%) on water surface ###*
1995	2.5	'92-'94	4110-18	27	No	5.4
1995	2.5	'92-'94	4110-18	27	Yes	2.7
1998	2.5	'97-'98	XR11004	20	No	2.9
1998	2.5	'97-'98	XR11004	20	Yes	0.6
2000	5	1998	DG1104+end	20	No	0.9
2000	3.3	1998	DG1104+end	20	Yes	0.15
2000	5	1998	ID12004	20	No	0.7
2000	3	1998	ID12004	20	Yes	0.15

##Spray drift deposition on water surfaces for potato groing in the Netherlands for 1995,1998, and 2000.

Note: ID series nozzles are air induction nozzles. Adapted from Zande, J.C. van de, (2002). Spray drift and buffer zones. Danske plantevaernkonference 2002. DJF rapport nr.66. pp69-71.

Sprayer Design

Air-sleeve booms are better at targeting the spray into the canopy, reducing drift and increasing deposition. Shields can vary from the simple to the complex, fitted with flat fan nozzles or controlled droplet applicators using reduced herbicide rates. Shielded sprayers allow growers to apply herbicides in variable weather conditions.

Trials were conducted by van de Zande (2002) over a number of seasons on potato crops. Their results, shows that air assistance reduced spray drift in all situations by 50% (sprayer boom height 27inches) to 70% (sprayer boom height 20 inches), independent of the nozzle being used. The results also show how nozzle technology has improved over the years, especially at reducing drift. Modern air induction nozzles reduce drift much better than traditional nozzles plus air assistance. The cost savings are also appreciable.

Sprayer Calibration

Correct calibration will ensure that all the nozzles are discharging the correct amount of liquid at the correct distance and angle to the target and at the correct forward speed.

Weather

Wind speed and direction, relative humidity, temperature and atmospheric stability affects drift. Applying the correct product to the correct target at the correct time with the correct equipment is the key to good spraying.

Shelterbelts

Research in England and New Zealand has been conducted to measure the effectiveness of shelterbelts. Natural and artificial belts were used and drift is reduced the closer you are to the shelterbelt. Shelterbelt height and density will affect drift, and may, in certain conditions, create additional air currents and eddies. There are so many variables such as topography and wind direction that it is difficult to conclude that research at one site is transferable to another.

Sprayer Size

Choose the correct size sprayer with good back-up support to ensure that spraying may be done in a timely manner. Far too often growers are racing around in an attempt to apply pesticides after a problem disease or insect attack has occurred. Good logistical support in reducing the need to return for frequent refills is so important. The use of field cards, detailing the block, pesticide required, application rate, quantity required per tank fill etc, will reduce stress levels found amongst some applicators.

Reducing Drift from Sprayers

Problem: Farmer Brown has a neighbour adjacent to his farm and would like to stop any drift from reaching his neighbours property.

Method

1. Monitoring equipment. Purchase and use good quality instruments for wind speed, temperature and humidity
2. Air Induction Nozzles (AI): These nozzles, when used properly, can reduce drift by at least 50 percent. The principle behind these nozzles is to create a larger droplet that won't drift as far but still maintain good leaf and fruit coverage. Note, not all AI nozzles are the same. Wilger and Lechler manufacture air-assist units to enable AI and hollow cone nozzles to be switched on/off from the tractor.
3. Shielded Sprayers: Shielded sprayers are the best way to reduce drift. Very little spray gets out of the shield spraying system allowing for a 90 percent reduction in drift. If A.I. nozzles are used within the shield sprayer 99 percent of drift can be reduced.

4. Drift reducing additives. A number of manufacturers supply drift reducing agents, most work via increasing droplet size. Beware not all of them can withstand the higher pressures associated with fruit sprayers and need independent verification.
5. Calibrate and check that sprayer is functioning correctly.

Drift is impossible to eliminate but can be minimised. Implementing just one of these methods will greatly reduce the effects of drift and improve your efficiency of spray application saving you time, money, and future problems.

Management Strategies to Reduce Drift

Before Spraying

1. Train the operator to use the sprayer correctly on your farm under your conditions.
2. Plan the spraying operation; consider the use of field work-cards, outlining the amount to be mixed/sprayed per block as a good management tool.
3. Read and follow the pesticide label.
4. Select the correct nozzle for the target. Adjust the size and position of the nozzles to achieve correct distribution within the canopy, particularly as the growing season progresses.
5. Consider the use of sprayers which direct the spray to the target such as canopy and air assist.
6. Consider spray additives to reduce drift only if they are independently proven to work.
7. Improve spraying logistics to ensure adequate time to spray within 'ideal' conditions.
8. Only spray when weather conditions are ideal; avoid spraying on days when conditions are favourable for atmospheric inversion or wind drift.
9. Calibrate the sprayer with water to ensure that everything is working correctly.
10. Start planting windbreaks!

During spraying:

1. Stay alert: ensure the spray is not allowed to drift on to non-target areas and watch for changes in wind speed and direction.

2. Keep spray pressure as low as possible and ensure an accurate gauge is used.
3. Maintain a constant speed and pressure, if an automatic regulator is fitted, remember, small increases in speed result in large increases in pressure. The delivered air and spray must be given time to penetrate the canopy.
4. Avoid spraying near sensitive crops or water courses. Use a 50-100 feet buffer zone; Spray inwards, with one side of the sprayer, for at least 50 feet from the boundary to create a "headland"

Table 2: Registration of BBA - Approved plant protection equipment for field crops in the list of loss reducing equipment

Drift Reduction Class	Sprayer or Sprayer Parts	Regulations of Use	
		Spray Pressure	Height above target
50%	Air Induction Nozzles: AVI, AirMix, S InJet, ID, MD, IDK, IDN, AI, AIC (various sizes)	43.5-101.5 PSI	20 in
	Twin Fluid Nozzles (various sizes)	43.5-72.5 PSI Air pressure 5 PSI	20 in
	HARDI Alpha or HARDI COMMANDER plus TWIN FORCE (59-92ft) all with HARDI ISO nozzles	Spray Pressure at 36 PSI fan drive at 2 KSI	12 in
75%	Air Induction Nozzles: AVI, AirMix, ID, MD, IDK, IDN, AI, AIC (various sizes)	14.5-43.5 PSI	20 in
	Dammann ANPA with air-assisted DAS(79-92ft) all with Air Induction nozzles: ID, AI, AirMix, AVI (various sizes)	Spray Pressure at 14.5-43.5 PSI fan drive at max 2.5 KSI	20 in
	HARDI Alpha or HARDI COMMANDER plus TWIN FORCE (59-92ft) all with HARDI ISO nozzles	Spray Pressure at 43.5 PSI fan drive at max	20 in
90%	Air Induction Nozzles: AirMix, S Injet, ID, MD, IDK, IDN (various sizes)	14.5-29 PSI	20 in
	Lechler ES	Band Spraying	

Decontaminating Boom Sprayers

Introduction

Sprayers must be thoroughly cleaned inside and out after use. Ideally, a sprayer should be cleaned at the end of each day and especially before switching to a different pesticide. Pesticide residues left on the outside of the sprayer can cause operator contamination. Residues on the inside of the tank or left over pesticides trapped inside the sprayer plumbing system can contaminate the operator and possibly lead to crop damage. Growers should be concerned about this, especially if they are using one sprayer to apply different chemicals to different crops. In some cases, only a small amount of a pesticide remaining in the sprayer can cause significant crop damage or lead to unacceptable residues on a crop.

Crop contamination can even occur several months after a sprayer has not been properly cleaned. Where a sprayer is used to spray different vegetable crops, residue left in the tank can cross contaminate another vegetable crop resulting in rejection by the purchaser.

Sprayers can also retain tremendous amounts of pesticide solution. Depending on the size and design of the sprayer, there can be nearly 6 gallons of solution left in a sprayer's plumbing. Research conducted on boom sprayers has shown that, depending on the spray tank size, the total chemical solution retained in the sprayer ranged from just under 3 gallons to over 12 gallons.

The parts that retained the most chemical solution are the chemical induction bowl, the booms, tank and the pump and its related piping. Tests have shown that triple rinsing the spray tank is better than using just one single rinse.

For example, using 100 gallons of clean water in one single rinse to clean a 100-gallon sprayer tank reduced the concentration of the original spray solution from 100% to 5% both in the tank and at the nozzle. If triple rinsing was performed using 33 gallons of clean water per rinse, a concentration of 0.2% to 0.5% was gained. The aim is for maximum dilution with minimal use of water.

Before rinsing a sprayer, read the sprayer manufacturer's instructions for specific guidance on the best methods for cleaning your equipment. Also consult the pesticide label for any special cleaning instructions. When cleaning spray equipment, you should use the protective clothing listed on the pesticide label. Sprayer cleaning should be done so that rinse water does not enter any waterway, field drainage system, or well.

Ideally, sprayer rinsate should be applied to a labeled crop rather than dumped at the cleaning location. If rinsing needs to be done at the mixing/loading site, it must be done on an impervious surface. All contaminated rinse water must be trapped and either used to mix another load of the same pesticide at the label recommended rates or disposed of at an approved pesticide waste handling facility.

Table 3: Quantity and location of chemical remnants in crop sprayer (in gallons)

	Sprayer Size		
Location	***159 gallons 39' boom***	***212 gallons 39' boom***	***396 gallons 59' boom***
Tank	0.50	1.32	4.57
Pump and associated piping	0.40	.85	2.22
Pressure agitation	0.02	.16	0.27
Manifold	0.40	.16	0.27
Filter relief valve	NA	.15	0.23
Chemical induction bowl	1.16	1.69	NA
Total without boom	2.12	4.33	7.56
Booms	0.50	2.32	4.76
Total with booms	2.62	6.65	12.32

Source: Nilsson, E., Hagenwall H. og Jorgensen L.

Table 4: Concentration of pesticide in rinse water

Rinse Number	***Sample Location***	***Percent Concentration***
1	Nozzle	5.5
	Tank Drain	4.8
2	Nozzle	1
	Tank Drain	1
3	Nozzle	0.2
	Tank Drain	0.2

Source: Nilsson, E., Hagenwall H. og Jorgensen L.

Reducing Cleaning Problems

The need for cleaning can be reduced by good planning and equipment maintenance. The following are suggestions to help reduce cleaning needs:

- Carefully plan how much pesticide to mix so that all mixed pesticides are used up when you are finished with the field.
- Be sure that the sprayer is clean before you use it.
- Make sure all parts of the sprayer are in good condition. Corroded, cavitated or pitted surfaces are prime areas for pesticide residue to hide. Replace any worn parts.

- Mix the chemicals in the correct order. Some chemicals, when mixed in the wrong order, can actually become more difficult to remove from the equipment. Consult the pesticide label for the proper mixing order.
- Follow any label instructions for cleaning spray equipment.
- Be sure that cleaning solutions contact ALL equipment surfaces.
- Remove and clean filters, strainers and nozzle screens separately from the rest of the sprayer.

Sprayer Cleansers

Several sprayer cleansers are commercially available. These cleansers should be selected based on the pesticide formulation used. Specific recommendations can be found on the pesticide label, by contacting the pesticide manufacturer or through the label or manufacturer of the cleaning agent you wish to use. Household detergents, such as laundry soaps and household ammonia, can also be used, but they may not adequately deactivate and solubalize the pesticides for effective cleaning. Chlorine bleach solutions should not be used. Cleaning agents can be used to wash both the inside and outside of the sprayer. When using commercial cleansers, follow the product's instructions for the best results.

Tank Rinse Systems (Low-Volume Tank Rinsing)

Tank rinse systems consist of a clean water supply tank mounted to the sprayer and one or more rotating discs or nozzles mounted inside the main sprayer tank. Water is pumped from the clean water tank to the rinse nozzles where the water is sprayed around the inside of the spray tank. These systems are designed for infield rinsing of the sprayer so that the tank washings can be applied to the field and reduce the amount of time spent travelling to and from the farmyard
.

A tank rinse system can be purchased as an option on some sprayers or as an add-on kit. Rinse systems can also be made from readily available parts and installed on the sprayer. A sample rinse system layout is shown below. A typical rinse system uses 360-degree tank wash nozzles mounted in the top of the tank. These nozzles are available in flow rates of 10 gallons of water per minute at 20 psi up to 20 GPM at 50 psi. If a spray tank has baffles, at least one rinse nozzle per compartment should be provided. In any case, a sufficient number of rinse nozzles should be installed to provide enough rinse water to contact the entire tank interior. A 50 to 100 gallon tank is plumbed into the sprayer plumbing system to provide the clean water. This tank should be permanently marked "Clean Water Only" so that only clean water is placed in the tank, reducing the chance for contamination of the rinse system. The tank should be mounted above the pump in order to aid in priming the pump. Ideally, the tank should be mounted on the sprayer.

When using tank rinse systems, you may want to check the pesticide label or with the chemical manufacturer to be sure that low-volume rinsing is suitable for the products you're using. Also, during he rinse process, be sure to open and close the pressure valve and other control valves on the sprayer to ensure that any chemical that may be trapped in the valve is rinsed out, further reducing the chance for contamination of future pesticide mixes. To obtain the best results, practice using the rinse system by placing spray marker dye or food colouring in the spray tank. Using the rinse system, run three rinse cycles, making sure the water discharged from the nozzles is completely clear by the end of the third rinse.

Cleaning the Sprayer

The pesticide applicator should try to keep the volume of tank wash water produced to a minimum. Ideally a tank rinse system should be used.

There are two levels of sprayer cleaning:

- where the same or similar products are to be used on consecutive occasions or
- where the type of product is changed for another or at the end of the season.

Cleaning Where Similar Products are to be Used

Reminder: Before cleaning application equipment, remember to wear the protective clothing listed on the pesticide label.

1. Be sure that all mixed pesticides have been used up from the sprayer or removed and disposed of properly.
2. Flush sprayer with clean water, making sure to wash all inside surfaces of the tank, including the underside of the lid. Use of a tank rinse system is preferred so that rinsing can be done in the field where the rinse water can be applied to the crop. If a tank rinse system is not available, fill the spray tank about half full with clean water and flush the system for at least 5 minutes using both agitation and spraying. Be sure to open

and close any control valves during the rinse process. The rinsate should be applied to the crop at labeled rates. Repeat this procedure two more times.

3. Hose down the outside of the sprayer making sure to reach all parts, scrubbing if necessary.
4. Remove suction, main and in-line filter elements and wash them thoroughly in clean water using a soft bristle brush. Put the filters back on the sprayer when clean.
5. Remove the nozzles, nozzle screens and nozzle bar end caps (if used) and wash them thoroughly in clean water with the appropriate cleanser and rinse. Remember to use a soft bristle brush, such as an old toothbrush, when cleaning nozzle parts.
6. Partly fill the sprayer with clean water and run the sprayer to flush out all parts.
7. Reinstall nozzles and nozzle screens.
8. Hose down the outside of the sprayer once again.

Cleaning Where Product Type is Changed

This procedure should also be followed at the end of a season or before sprayer maintenance. *Reminder*: Remember to wear the protective clothing listed on the pesticide label.

1. Be sure that all mixed pesticides have been used up from the sprayer or removed and disposed of properly.
2. Flush sprayer with clean water, making sure to wash all inside surfaces of the tank, including the underside of the lid. Use of a tank rinse system is preferred so that rinsing can be done in the field where the rinse water can be applied to the crop. If a tank rinse system is not available, fill the spray tank about half full with clean water and flush the system for at least 5 minutes using both agitation and spraying. Be sure to open and close any control valves during the rinse process. The rinsate should be applied to the crop at labeled rates. Repeat this procedure two more times.
3. Hose down the outside of the sprayer making sure to reach all parts, scrubbing if necessary.
4. Remove suction, main and in-line filter elements and wash them thoroughly in clean water using a soft bristle brush. Put the filters back on the sprayer when clean.

5. Remove the nozzles, nozzle screens and nozzle bar end caps (if used) and wash them thoroughly in clean water with the appropriate cleanser and rinse. Remember to use a soft bristle brush, such as an old toothbrush, when cleaning nozzle parts.
6. Partly fill the sprayer with clean water and run the sprayer to flush out all parts.
7. Refill the tank with clean water, adding any detergent recommended by the pesticide manufacturer. Remember, use commercial cleansers according to their directions. Agitate the solution and pump it through the sprayer plumbing system.
8. Discharge the cleaning solution from the sprayer through the plumbing system, making sure to drain the system as thoroughly as possible.
9. Rinse the sprayer and flush the plumbing system with clean water.
10. Inspect the sprayer for deposits that may remain in the tank or plumbing system. If any remain, use some of the cleaning solution and scrub the problem spots. Rinse the sprayer out completely.
11. Repeat steps 7 to 9.
12. Hose down the outside of the tractor and sprayer, scrubbing if necessary.
13. If changing from one type of pesticide to another, refit nozzles, filters and other parts that may have been removed in the cleaning process.
14. When cleaning and preparing the sprayer at the end of the season, safely store nozzles and filters to keep them clean and damage-free. Leave valves open and the tank lid loosely closed.

Mechanical Maintenance

Lubrication must be carried out prior to storage, check all oil levels. Check the soundness of all mechanical components, particularly booms and boom hinges. Electrical connectors which operate boom control valves, spray monitors etc need to be cleaned and a non-conductive grease, available at an auto store, applied to prevent corrosion. Check wheels, wheel bearings and tire inflation.

Storage of Sprayers

Store the sprayer under cover, taking care to prevent dirt and moisture affecting the tank or working parts. Remember, sunlight softens and weakens rubber materials and can degrade plastic materials. Storing in a building also allows you the opportunity to conduct any routine or pre-season maintenance.

Spray Additives

The word 'adjuvant' is used for any chemical that, when added to a liquid spray, makes it mix, wet, spread, stick, or penetrate better. Water is almost a universal diluent for pesticide sprays. However, water is not compatible with oily pesticides, and an emulsifier may be needed to obtain good mixing.

Furthermore, water from sprays often remains as large droplets on leaf surfaces. A wetting agent lowers the interfacial tension between the spray droplet and the leaf surface and thus moistens the leaf. Spreaders are closely related to wetters and help to build a deposit on the leaf and improve weatherability.

Stickers cause pesticides to adhere to the sprayed surface and are often called spray-stickers. They are oily and serve to increase the amounts of suspended solids held on the leaves or fruits by holding the particles in a resin-like film. Most fungicides greatly benefit from the use of a sticker. A new group of materials, called extenders, has been developed recently. Extenders form a sticky, elastic film that holds the pesticide on the leaves and thus reduces the rate of loss caused by sunlight and rainfall.

There are a number of adjuvants on the market including Triton B 1956, DuPont SS, Ortho SS, Plyac, Triton CS7, Nu Film 17, Pinolene, Ortho X77, Sun llE, Booster + E, and others. Probably no two are identical, and some may be quite specific for crop or pesticide use. Read the label, both for dosages and for crop uses and compatibilities, since some adjuvants must not be used with certain pesticides. Most labels will suggest amounts needed. Although many formulations of pesticides contain adequate adjuvants, some do require additions on certain crops, especially cabbage, cauliflower, onions, and peppers.

Spray adjuvants for use with herbicides often serve a funcon distinctly different from that of adjuvants used with insecticides and fungicides. For example, adjuvants such as oils used with atrazine* greatly improve penetration of the chemical into crop and weed leaves, rather than just giving more uniform coverage. Do not use any adjuvant with herbicides unless there is a specific recommendation for its use, either on the label or from extension personnel. Plant damage or even crop residues can result from using an adjuvant that is not recommended.

Spray Equivalents, Conversions, and Dosages

Pesticide containers give directions in terms of pounds or gallons of material in 100 gallons of water. Recommendations for rates of pesticides are often stated in pounds of active ingredient per acre. However, commercial products are not 100 percent active.

They have been specially formulated by adding ingredients such as solvents and adjuvants to improve the sprayability, storage life, and other properties of the pure chemical. The additions reduce their concentration.

For example, dry concentrates contain perhaps 50 percent or 75 percent active ingredients; liquid concentrates, 50 percent or four pounds per gallon. Thus, to obtain one pound active ingredient, two pounds of formulation may be needed. Refer to the label to determine correct amounts.

Storage

Pesticide Storage

Most pesticides will retain their effectiveness for more than a year if properly stored. Low temperatures, excessive heat and/or inadequate storage conditions can ruin pesticides. Always store pesticides in their original containers with the labels securely attached and plainly visible. If material is to be stored for a long period, you can protect the labels by covering them with transparent tape or lacquer. Mark each container with a purchase date and keep a complete inventory of all products. Keep this list readily accessible in a separate building. Do not store herbicides with other pesticides since some herbicides are volatile and readily contaminate other chemicals.

Table 5: Signs of deterioration in pesticides

Formulation	General Sign of Deterioration
Wettable powders	Excessive lumping or caking; powder will not stay in suspension.
Oils	Milky colouration or clouding does not occur when added to water.
Liquid concentrates	Crystals form, separation or layering of components; sludge is present, milky colouration does not occur when added to water.
Granular	Excessive lumping or caking into useless blocks.
Flowables	Settling out or layering of components.

Pesticide Storage Building

Site

- All-weather accessible site - loading, unloading, fire brigade access
- 50 feet from other buildings
- 100 feet min. from wells/water supplies 7 down slope from them .

Building

Comprise of:

- Pesticide storage room.
  * Mixing room
  * Locker room
  * Filling/loading pad
  * Electrical power for lights, ventilation fans and heatin
  * Sign to say pesticide storage, Danger etc.

Pesticide Storage Room

- Well ventilated, perhaps electric fan fitted via light switch to ensure rapid ventilation
- Lights to see the labels
- Shelves; heavy drums on lower shelves, lighter packs above
- Base should be a bund to prevent seepage, to contain fire water
- Roof: insulated but light to blow out in fire/explosion
- Keep all containers off the floor, eg. pallets
- Locked door
- Contents list
- Rotate stock, oldest at the front.

Mixing Room

- Keep all calibration equipment, all measuring jugs, records etc., on organised shelves.

Locker Room

- PPE, ventilated, check list of issue dates
- Definitely not in the pesticide storage room
- A cupboard perhaps in filling /load area.

Filling/Loading Pad

- Water supply: one way (anti-siphon) valve to prevent chemicals into the water supply
- Consider a top hat cage to provide a gap between hose end and tank top
- Water pipe: insulate
- Eye wash unit
- Major dousing hose or hook up filling hose to become a dousing site

Indoor versus Covered

Outdoor: Cheaper but Frost Heave and Rain Water Collects

Mixing room and filling pad should also hold any spillage, e.g. Sprayer tank contents plus 25%; Sump or sloping concrete

Construction

- Insulation of R11 on outside walls, 4" fibreglass
- R19 in the ceiling, 6" fibreglass
- Air inlet with louvers for the opposite of the fans
- Beware of draughts, therefore fans/inlets above top storage shelves
- Plastic sheet under concrete: keep building dry
- Gutter down pipe: keep intact and water away from the building
- Electrically shielded fan, some suggest shielded lights
- Fire extinguisher.

Chapter 8

Soil Fertility and Biodiversity in Organic Farming

Fertility management is part of an overall soil management program that involves proper tillage practices, crop rotation, cover crops, water management (irrigation and drainage), liming, and weed management. Although it is important in obtaining maximum economic yields, fertilization alone will not overcome shortcomings in the areas mentioned above. Such problems should be corrected first so as to benefit fully from organic and inorganic fertilizer supplements and to sustain high yields and quality over the long term.

Plan ahead when selecting new lands or fields. Soils for growing vegetables should be well drained, fairly deep, reasonably level, properly limed, and in good tilth (have good structure). Medium-textured soils (sandy to silty loams with good organic-matter content) are generally most satisfactory; well-drained, sandy soils with a slight to moderate southern slope are most favourable for early plantings and certain warm-season vegetables. For a summary of soil types and soil management groups in New York State. Detailed soil survey maps are available through local extension and NRCS offices. After determining whether the soil is suitable, check for perennial weeds and correct pH level before planting.

Soils in good tilth provide a desirable medium for root development, absorb water and air rapidly, withstand erosion, and resist the tendency to crust. Many agricultural practices cause soil structure to deteriorate. Compaction, which results from the use of heavy equipment on wet soils, is particularly damaging. Tillage tools break down soil aggregation; intensive cultivation accelerates loss of organic matter and causes soil to crust. Obviously, all unnecessary operations should be avoided. Prepare the soil only enough to provide an adequate seedbed. Never plow, till, plant, or cultivate soils when they are wet.

One mistake can reduce the yield of the planted crop regardless of the level of other inputs.

Cover Crops

Cover crops are close-growing crops planted primarily for protecting and improving the soil (photos in NY). Integrating cover crops into vegetable production systems offers many benefits, but provides some challenges as well. For cover cropping to be successful, it is important to know the intended purposes, consider key management factors, and understand the characteristics of different cover crop species.

Cover crops offer a way to add organic matter to soils; improve soil tilth and remediate compaction; protect soil from wind and water erosion; add or recycle plant nutrients; increase the biological activity of soil; retain soil moisture; and in some cases, suppress weeds and other pests. No single cover crop can do all of these things. Matching the need and opportunity to the right cover crop requires information and planning.

Cover crops will perform best under good growing conditions, such as optimal temperatures, sufficient soil moisture, and adequate soil fertility. Practices, such as preparing an adequate seedbed; drilling seed or broadcasting and cultipacking; inoculating seed with the proper *Rhizobium* inoculant if using a legume; planting into sufficient soil moisture; correcting pH or soil fertility problems; and in some cases, controlling weeds with herbicides or mowing the cover crop in midseason, often further enhance cover crop performance. Access to appropriate equipment for incorporating the cover crop is also critical.

Despite the relatively short growing season in New York, there are many ways to integrate cover crops into vegetable rotations. Examples include planting grain rye or wheat after vegetable crop harvest; interseeding ryegrass or clover into a standing vegetable crop; no-till planting pumpkins into a herbicide-killed rye mulch; planting barley windbreaks in muck-grown onions; or taking land out of production for a season.

The costs associated with cover cropping, which include the price of seed, amount of time and labour invested, use of equipment, and value of missed opportunities, should be taken into account. Some cover crops pose specific management challenges. For example, either grain rye left to grow into late spring or unmowed Sudan grass left into the summer can result in such rank growth that they are difficult to incorporate into the soil and may cause nitrogen immobilisation. Cover crops can sometimes have detrimental effects. Direct-seeding a vegetable crop into soil with fresh organic residues invites seed-maggot damage. Some cover crops, such as hairy vetch, are hosts for common nematodes or soilborne diseases such as white mold. Before growing any new cover crop, try it first in small test strips.

Manure

Most vegetable operations do not have a ready source of manure, but it should be used when available. Once applied to soil, manure is decomposed by microorganisms, forming humus. Manure provides both major and minor nutrients, and when used regularly, it contributes organic matter and helps to alleviate structural deterioration, an important consideration in maintaining the productivity of heavily worked vegetable soils. One drawback of using manure is that certain weed seeds maintain their viability after passage through animals, so a potential exists for adding a new weed species to a field. This threat is more likely with fresh than with composted manure. An excellent, thorough discussion of manure use in crop production is provided in the *Cornell Field Crops and Soils Handbook*.

Manure contains two forms of nitrogen, the unstable form in the urine and the stable form in the feces. The unstable form may account for 50 percent or more of the total nitrogen in manure. This nitrogen decomposes rapidly to ammonium, which in turn converts quickly to extremely volatile ammonia that can be lost from the system. For this reason, much of manure's unstable nitrogen may never be taken up by crops unless measures are taken to conserve it during the process of collection, storage, and application to the field. In general, about 35 percent of the stable nitrogen becomes available during the year of application, about 12 percent the second year, about five percent the third year, and about two percent the fourth year. Thus, repeated application to the same field results in an accumulation of a slow-release source of manure nitrogen.

Most potassium in manure is available for plant growth during the year applied; whereas, some of the phosphorus is in organic form and must decompose before it becomes available. Moreover, because phosphorus is not very mobile in the soil, broadcasting manure is not an efficient way of applying this element for crop establishment.

A micronutrient deficiency in a field with a history of manuring is rare because manure contains small quantities of these elements. If a deficiency is observed on a nonmanured field, a commercial fertilizer should be added immediately because of the slower availability of micronutrients in manure. If soil pH is acceptable, manuring may eventually solve the problem.

Manure and Produce Safety

The use of improperly aged or treated manure can increase microbial risks and contribute to foodborne illness. The possibility that fecal matter may come into contact with produce or that water might splash pathogens from the manure onto field produce are both important concerns. Pathogens such as *E. coli* O157:H7, *Salmonella*, and *Campylobacter* can be present in manure slurry for up to 3 months or more, depending on temperatures and soil conditions. Troubling for growers is that *Listeria monocytogenes* can survive in the soil for much longer than 3 months. *Yersinia enterocolitica* may survive, but not grow, in soil for almost a year.

It is important that all farms using manure follow good agricultural practices to reduce any microbial risk that may exist. These include;

Consider the source, storage, and type of manure

- Store manure as far away as practical from areas where fresh produce is being grown and handled. If manure is not composted, age the manure at least six months prior to field application.
- Where possible, erect physical barriers or wind barriers to prevent runoff and wind drift of manure.
- Store manure slurry for at least 60 days in the summer and 90 days in the winter before applying to fields.
- Compost manure. High temperatures achieved by a well-managed, aerobic compost can kill most harmful pathogens. Remember to optimise temperature, turning, and time to produce high quality, stable compost.

Plan Manure Application Timing Carefully

- Apply manure in the fall or at the end of the season to all planned vegetable ground, preferably when soils are warm, nonsaturated, and cover cropped.
- Incorporate manure into soil.
- If applying manure in the spring, spread the manure two weeks prior to planting, preferably to grain or forage crops.
- DO NOT harvest vegetables until 120 days after manure application. If this is not feasible for short season crops like

lettuce and leafy greens, apply only properly composted manure.

- Document rates, dates, and locations of manure applications.

Choose Appropriate Crops

- Avoid growing root or leafy crops in the year that manure is applied to a field
- Apply manure to grain or forage crops or to vegetables that will be heat processed
- Apply manure to perennial crops in the planting year or immediately following the last harvest of the season.

Sewage Sludges

Sewage sludges are a by-product of the purification of wastewater. Sewage sludges have a significant organic matter content and contain micro- and macronutrients essential for plant growth. Sewage sludges also contain contaminants such as heavy metals, toxic organics and pathogens. State and federal rules require that before being applied to agricultural lands, sludges must undergo further treatment for stabilisation and disinfection. Restrictions on use depend on the level of treatment. Prior to using sludges, growers should carefully review the regulations. Many food processors will not accept vegetables grown on sludge-amended soils, so growers should check before applying sludges. Consumers may also be unwilling to accept such crops.

The quality of sewage sludges varies widely. Growers who are considering sludge use are urged to obtain analytic information from the supplier on the metal, toxic organic and nutrient concentrations in the particular sludge or sludge product they might apply.

Regulations require testing only for a very limited list of regulated metals, nutrients and pathogens and only at intervals determined by the size of the sewage plant (infrequent testing at smaller plants). Testing for additional (unregulated) contaminants including antimony, beryllium, boron, chromium, molybdenum, selenium, silver, sulfur, and for dioxins and PCBs is recommended, as is testing of the particular load(s) to be applied, because composition and water contents may vary significantly.

Application rates are generally calculated on the basis of nitrogen content and crop N requirements. The nitrogen mineralisation rate of sludges is not well characterised and the nitrogen content of sludge from even one particular treatment plant varies widely, making it difficult to determine appropriate application rates.

Although many studies have been done on the land application of sludges, scientific uncertainties remain, particularly with respect to long-term soil productivity and ecologic safety. Other potential concerns include odours, airborne toxins and pathogens, and ground and surface water impacts.

Cornell Soil Health Website and Manual

Soil Testing

Fertilizer requirements for best economic yield should approximate the difference between what vegetables take up from the soil for best growth and quality and what the soil can actually supply during the crop-growing period. The supply of essential nutrients in soil cannot be determined without conducting a soil test. Moreover, if pH is not in a desirable range, yields may be poor regardless of fertilizer added or already present in the soil.

Soils on which vegetables will be grown should be sampled at least once every three years. The pH of most vegetable soils decreases gradually because of the removal of bases by leaching and crop uptake and from acid forming fertilizers. Testing every year gives a more complete evaluation and is appropriate when significant changes have been made in the fertilizer program (e.g., applying less phosphorus or potassium when the previous year's test showed high levels). In general, when the Cornell-recommended rates of fertilizer are applied, low soil test values for phosphorus and potassium usually increase slowly and steadily in spite of crop removal. Medium soil test values tend to remain constant or increase slightly, whereas high values decrease gradually. The potassium level could decrease much more rapidly, however, if a light sandy soil with relatively low exchange capacity is coupled with a heavy potassium feeder such as potatoes or tomatoes. In such situations, yearly sampling is appropriate. The purpose of applying nutrients, however, is to benefit crop development, not to achieve a predetermined test result.

Growers interested in obtaining Cornell guidelines for crop management should submit samples to AgroOne and request the Cornell Morgan test and Cornell guidelines (730 Warren Road Ithaca NY 14850; 1-800-344-2697). For New York samples submitted with the necessary field information, the soil test report from AgroOne will contain recommendations that are derived from the Cornell fertility guidelines engine (software system) using Cornell Morgan equivalents as inputs.

The soil test results provide soil pH, percent of organic matter, and level of phosphorus, potassium, magnesium, calcium, and zinc. Levels of aluminium, iron, and manganese are also listed to identify potential toxicities rather than deficiencies. Other nutrients can be tested for an additional fee.

Cornell Soil Health Website and Manual

Soil pH

In general, vegetable crops grown on mineral soils will thrive at pH 6.0 to 6.5. Some vegetables do well at pH 5.5; potatoes will tolerate even greater acidity. In contrast to mineral soils, the desirable pH for muck soils is approximately 5.5, and they should not be limed above pH 5.7. This is largely because of the much greater amounts of calcium found in muck at pH 5.5 compared to mineral soils at similar pH. Specific pH ranges for individual vegetable crops are given under each crop's fertility section.

When soil pH is adequate, the availability of both major and minor nutrients is maximised, and the accumulation of toxic metals is minimised. Clearly, one cannot expect to maximise dollars spent for nitrogen, phosphorus, and potassium fertilizer when soil pH is suboptimal. Thus, many people consider soil pH to be the most important part of the soil test.

For optimal vegetable production on New York's acidic soils, soil pH should be adjusted with lime to fit the needs of the various crops. When soil pH is 6.0 or below, the laboratory will determine the exchangeable acidity on the sample. A lime requirement can be determined based on pH (actual acidity of the soil solution) and exchangeable acidity (reserve acidity for that soil). Two soils with the same low pH reading could require markedly different amounts of lime to correct the situation. A given amount of lime could overlime one soil, causing problems that did not exist previously, whereas the same amount of lime might be insufficient to correct the undesirable acidity in the other soil. This is one reason soil testing is so important.

Based on exchangeable acidity, as determined by the soil test, an accurate lime recommendation can be given. When complete soil tests are not available. The lime rates given are based on an eight inch plow depth. If the plow depth is less than six inches, decrease the rate given in the tables by 12 percent for every inch less than eight inches. If the plow depth is more than eight inches, increase the lime rate given by 12 percent for every inch greater than eight inches. For example, a plow depth of ten inches and a lime recommendation of four tons would require 24 percent more lime than given in the table. Therefore, the total rate to apply is approximately five tons (4 tons multiplied by 1.24 = 4.96).

These rates need to be increased or decreased according to the actual E.N.V. of the limestone being applied. The rate to be applied is calculated by dividing the recommended rate given in the tables or the test report (if necessary, correct for plowing depth) by the E.N.V. of the lime to be used. For example, if the recommended lime rate is four tons per acre and the E.N.V. of the limestone to be spread is 0.68, the rate to apply would be 5.88 tons (4.0 divided by 0.68), which would round off to six tons per acre. The delivery slip accompanying bulk spread limestone specifies the E.N.V. and the quantity required to equal limestone at 100 percent E.N.V.

Limestones vary in E.N.V. because of differences in purity (calcium carbonate equivalence) and particle size. Coarse limestone or limestone of lower purity is less effective than is more finely ground limestone or limestone of higher purity in neutralising soil acidity. Cost per ton can be misleading if the limestones being compared do not have a similar E.N.V.. Accordingly, the least expensive lime in terms of dollars per ton may not be the best value.

Limestones are mixtures of calcium carbonate and magnesium carbonate with calcium carbonate predominant. The magnesium content in limestones sold in New York varies from 1/5 percent to more than 12 percent. A dolomitic limestone contains a relatively high percentage of magnesium, although there is no legal definition. If pH is low, the magnesium soil test is low (below 65 pounds per acre), and a magnesium-sensitive crop such as melons or tomatoes is to be grown, using a dolomitic limestone is an excellent, economical way to provide the needed magnesium. If the soil test magnesium level exceeds 100 pounds per acre, there is no particular advantage to using lime with higher magnesium content. When magnesium levels are high, either a calcitic- or dolomitic-type limestone is appropriate.

If pH is below 5.5 on mineral soil, lime should be applied long enough before seeding the more acid-sensitive vegetables to react with the entire plow layer. If there is insufficient time for an adequate reaction with the entire plow layer, a split application is recommended. At least half of the recommended lime should be added to the surface and disked in before seeding to provide a pH favourable for good seedling establishment in the zone near the seed. When soils require more than four tons per acre, split the lime application by plowing one-half down and disking the remainder into the surface. Smaller lime applications to maintain pH above 6.0 can be made anytime before seeding and can either be applied to the surface or plowed down. When rotations are used, the last summer or fall that a field is in sod is a good time to apply smaller maintenance applications of lime. At this time the soil is firm, and lime can be applied with less likelihood of soil compaction.

Table 1: General lime recommendations to raise pH to 6.5 for vegetables other than potatoes.

Initial soil pH	*Sands*	*Sandy loams*	*Loams and silt loams*	*Silty clay loams*
		(Values given in Tons/acre)		
4.5	4.0	7.0	12.1	15.0
4.6 to 4.7	3.5	6.5	10.0	13.0
4.8 to 4.9	3.0	6.0	9.0	12.5
5.0 to 5.1	2.5	5.5	8.5	12.0
5.2 to 5.3	2.0	4.5	6.5	8.0
5.4 to 5.5	1.5	3.0	4.0	6.0
5.6 to 5.7	1.0	2.0	3.0	5.5
5.8 to 5.9	0.8	1.8	2.5	3.5
6.0 to 6.1	0.6	1.5	2.0	3.0
6.2 to 6.3	0.5	1.0	1.5	2.5
6.4 to 6.5	0.3	0.8	1.3	2.0

1: A guideline for muck is that 1,000 lb. of lime will raise the pH about 0.1. These rates are based on an 8" plow depth and lime with an effective neutralising value (E.N.V.) of 100 percent.

Cornell Soil Health Website and Manual

Fertilizers

Nitrogen, Phosphorus, and Potassium: Although necessary for high-yielding crop production, fertilizer nutrients can escape from the agricultural system, thereby increasing the potential for environmental damage. Nutrients escape in various ways depending on the chemical and biological nature of the element involved. Obviously, this escape can be accelerated if fertilizers are added in excess of plant requirements or if they are applied or otherwise handled improperly.

Regardless of the chemical form added, nitrogen can convert rapidly to nitrate; in this form it does not bind to the soil but rather moves downward with water as the water moves through the soil. Thus, excessive nitrate-nitrogen poses a threat to the quality of well water and ground water. Nitrogen is also lost to surface water as soils erode, removing organic soil matter.

Using nitrogen efficiently is probably the biggest challenge in fertility management. Vegetables are responsive to nitrogen, and no one wants to risk inferior yield or quality because of a deficiency. However, it is difficult to accurately determine the nitrogen contributions from soil organic matter, manure, or incorporated legumes because temperature and moisture can play a significant role. Also, sources of nitrogen convert to nitrate-nitrogen when conditions are optimal for plant growth, and in this form nitrogen moves with water and can be leached out of the system.

Guidelines for efficient use of nitrogen

1. Limit nitrogen applications prior to planting, and avoid deep plow-downs.
2. Band either at planting or as a sidedressing to apply nitrogen most efficiently.
3. Apply nitrogen close to the time the crop is most active in taking it up.
4. Avoid "insurance" applications of nitrogen.
5. Maintain the proper pH.
6. Use plastic mulch to limit leaching and facilitate nitrogen release from nonfertilizer sources.
7. Avoid over fertilization which will lead to leaching.
8. Account for nitrogen from organic matter, cover crops, composts, manure, etc., which becomes available as the soils warm.
9. Consider using the pre-sidedress soil nitrate test (PSNT) to determine nitrogen contributions from nonfertilizer sources.
10. Use cover crops to retain nitrogen and other nutrients and limit leaching.

Phosphorus is usually tightly bound to soil particles with only small amounts in the soil water. Phosphorus may also occur in organic soil materials, some of which are water soluble. Most phosphorus loss is attributable to surface runoff and soil erosion. Techniques that help prevent nutrient loss to the environment include prevention of soil erosion, avoidance of overfertilization or insurance applications, and timing and placement of fertilizer applications in a manner to achieve efficient plant uptake.

Fertilizers are applied to improve plant growth by providing nutrients not adequately supplied by the soil. When the soil contains enough of a particular nutrient to support optimal plant growth, there is no need to supply additional quantities of that nutrient. The most common nutrients in commercial fertilizers are nitrogen (N), phosphorus (P), and potassium (K). Phosphorus and potassium are shown on fertilizer labels as the oxides P_2O_5 and K_2O, respectively. For conversion multiply P_2O_5 by 0.44 to get P, and multiply K_2O by 0.83 to get K. Calcium (Ca) and magnesium (Mg) are usually supplied by liming.

Table 2: Common fertilizer materials supplying primary nutrients.

Common name	*Analysis Percent Chemical formula*	*N*	*P2O5*	*K2O*
Nitrogen materials				
Ammonium nitrate	NH_4NO_3	34	0	0
Ammonium sulfate	$(NH_4)_2SO_4$	21	0	0
Ammonium nitrate-urea[1]	NH_4NO_3+$(NH_2)_2CO$	32	0	0
Anhydrous ammonia	NH_3	82	0	0
Aqua ammonia[1]	NH_4OH	30	0	0
Calcium nitrate	$Ca(NO_3)_2$	15.5	0	0
Sodium nitrate	$NaNO_3$	16	0	0
Urea	$(NH_2)_2CO$	46	0	0
Phosphate materials				
Ordinary superphosphate	$Ca(H_2PO_4)_2$	0	20	0
Concentrated superphosphate	$Ca(H_2PO_4)_2$	0	45	0
Ammoniated superphosphate[1]	$Ca(NH_4H_2PO_4)_2$	54	0	0
Monoammonium phosphate[1]	$NH_4H_2PO_4$	13	52	0
Diammonium phosphate[1]	$(NH_4)_2HPO_4$	28	46	0
Urea-ammonium phosphate	$(NH_2)_2CO$+$(NH_4)_2HPO_4$	28	28	0
Potash materials				
Muriate of potash	KCl	0	0	60
Sulfate of potash	K_2SO_4	0	0	50
Potassium nitrate	KNO_3	13	0	44
Monopotassium phosphate[1]	KH_2PO_4	0	50	40
Sulfate of potash-magnesia (11% Mg)	$K_2SO_4 \cdot MgSO_4$	0	0	22

Table 3: Fertilizer materials supplying secondary and micronutrients.

Common name	Element	Chemical formula	Analysis %
Secondary nutrients			
Calcium nitrate	calcium (Ca), N	$Ca(NO_3)_2$	22 Ca, 15.5 N
Magnesium sulfate [1]	magnesium (Mg)	$MgSO_4$	16 Mg, 14 S
Magnesium oxide[1]	magnesium (Mg)	MgO	45 Mg
Superphosphate	sulfur (S)	$Ca(H_2PO_4)_2+CaSO_4$	8 S
Calcium sulfate (gypsum)[1]	sulfur (S)	$CaSO_4$	15 S
Sulfate of potash-magnesia	sulfur (S), Mg	$K_2SO_4\ MgSO_4$	22 S, 11 Mg
Potassium sulfate	sulfur (S)	$(NH_4)_2SO_4$	24 S
Micronutrients			
Borax[1]	boron (B)	Na_2BO_4	12 B
Solubor	boron (B)	Na_2BO_4	20 B
Manganese sulfate[1]	manganese (Mn)	$MnSO_4$	28 Mn
Zinc sulfate[1]	zinc (Zn)	$ZnSO_4$	36 Zn
Zinc oxide[1]	zinc (Zn)	ZnO	50 Zn
Zinc chelate[1]	zinc (Zn)	Zn-chelate	14 Zn
Ferrous sulfate[1]	iron (Fe)	$FeSO_4$	20 Fe
Iron chelate[1]	iron (Fe)	Fe-chelate	14 Fe
Copper sulfate	copper (Cu)	$CuSO_4$	25 Cu
Sodium molybdate	molybdenum (Mo)	Na_2MoO_4	39 Mo
Ammonium molybdate	molybdenum (Mo)	$(NH_4)_6Mo_7O_{24}$	54 Mo

Secondary Nutrients and Micronutrients

The secondary nutrients - calcium (Ca), magnesium (Mg), and sulfur (S) - are as important for normal growth as the primary nutrients but either are not required in large quantities or are usually supplied through means other than fertilizers. Micronutrients, often referred to as minor elements, include boron, zinc, manganese, copper, molybdenum, and iron. They are as important to normal plant growth and reproduction as are the primary and secondary elements. The difference is that micronutrients are required in small amounts because crops remove less than a pound per acre (less than an ounce per acre of some elements). Micronutrients are seldom deficient in New York soils when the pH is between 6.0 and 6.8 on upland soil and between 5.4 and 6.0 on muck. Response to micronutrients is rare on upland soils of reasonable organic matter content or on manured soils whose pH is in the proper range. In general, micronutrients should not be included in the fertilizer program. In a few specific cases, micronutrients may need to be added to achieve maximum marketable yield.

Fertilizer Placement

To be used by plants, nutrients must be present in moist soil where roots are active. Nitrogen placed on the surface will leach down to roots if irrigation or rainfall is adequate. Phosphorus and potassium do not move extensively and therefore should be placed deeply enough to remain in moist soil throughout the growing season. All of the phosphorus and potassium and some of the nitrogen should be applied shortly before or at planting; the remaining nitrogen should be applied as sidedressings during the early part of the growing period. Band application of at least part of the fertilizer near the seed or plant row is recommended, especially when soil test levels are marginal and crop response to the fertilizer nutrients is likely. Beware of banding too close with rates above 80 to 100 pounds per acre of potassium and nitrogen combined because seeds or seedlings can be injured.

Fertilizer Injury

All nitrogen and potash materials add water-soluble ions to the soil. If the concentration of these ions is too high near the germinating seed or seedling, salt burn can result, reducing germination and retarding seedling growth. The problem occurs most often when the weather turns dry after seeding, a salt-sensitive vegetable such as beans is being grown, or the fertilizer band is being placed closer to the seed than the grower intended. To prevent salt burn when banding fertilizer, avoid using more than 80 to 100 pounds of N + K_2O per acre in the band at planting. Potash can be broadcast and incorporated separately. This rule applies to fertilizer bands placed two inches below and two inches to the side of the seed. Check equipment when banding to ensure that the band is being placed where intended. If more than 80 to 100 pounds of N + K_2O per acre will be used, the band should be moved three inches to the side of the seed.

Materials containing nitrogen produce another type of germination or seedling injury associated with a high concentration of ammonia. Fertilizers producing this injury contain urea, diammonium phosphate (DAP), or anhydrous ammonia. Exceeding 30 pounds of nitrogen as urea, 30 pounds of phosphorus as DAP, or 30 pounds combined from fertilizers containing both materials may cause seedling injury in bands, especially when dry weather follows planting and the band is closer than two inches. Both urea and DAP can be used for plow-down applications without concern for injury. If anhydrous ammonia will be used as a preplant or preemergence source of nitrogen for sweet corn, it should be injected as far as possible from the seed.

Fertilizer/Transplant Solution

Adding a small amount of water-soluble or liquid fertilizer to transplant water can stimulate growth of young transplants such as cabbage, tomatoes, and peppers. Many grades are available for this purpose (e.g., 10-52-17, 14-28-14, 23-21-17, 20-20-20, 6-24-6, and 10-34-0). They are generally used at a concentration of about three pounds per 50 gallons of water and about one-third this strength on melon and cucumber plants. Transplants can be injured in hot weather if the soil is relatively dry. Reducing the concentration of starter fertilizer may help, but it is safer to irrigate before or immediately after transplanting.

Response to starter solutions is most likely when soils are cooler and tests indicate low phosphorus and potassium. In tests at Cornell with transplanted tomatoes under these conditions, diluted 10-34-0 liquid fertilizer was as effective as complete grades. Tomatoes transplanted later (early June), however, into a warmer identical soil testing high in phosphorus and potassium did not respond to 10-34-0 or a complete starter fertilizer. If container-grown plants have been fertilized just before transplanting, the starter fertilizer may be diluted or eliminated from the transplant water.

Fertigation

The most efficient way to fertilize an established mulch row crop is through a trickle irrigation system which is usually installed during the mulching operation. Due to the small holes in the trickle tubing, it is important that only completely soluble fertilizers are used. Best results have been achieved by using a 1-1-1 ratio of completely soluble fertilizer such as 20-20-20.

When applying soluble fertilizer, the system should first be fully charged with water. After the fertilizer has been injected, water should be applied only long enough to flush all of the fertilizer through the lines. A long irrigation immediately after fertigating will cause leaching of the fertilizer below the root zone and reduce, rather than increase, the efficiency of fertilizer. Drip fertigation schedules vary with the crop, soil conditions, and management practices. In general, preplant fertilizer should supply about 20 to 30 percent of the nitrogen and potassium and all of the phosphorous needs of the crop. The remainder of the nitrogen and potassium is supplied through the system during the growing season as needed by the crop. If equipment is available, applying the preplant fertilizer to the soil area to be covered with mulch is more efficient than a broadcast application. Some growers, particularly in initial years of experimenting, may supply up to 50 percent of the nitrogen and most of the potassium before planting.

To calculate fertilizer rates for trickle irrigation under mulch, base the amount applied on mulched acres rather than actual acres. For example, if the soil surface covered by the mulch is three feet wide and the row centre is six feet wide, you should apply 1/2 or 50 percent of the rate that would have been calculated for broadcasting on a per-field-acre basis. This is similar to reducing application amounts when banding fertilizers or herbicides.

The first soluble fertilizer should be applied through the drip system within one week of transplanting. The remainder should be split between three and five additional applications. Heavier soil may require fewer applications. There is little advantage in applying fertilizer more than once a week or once every other week, except on sandy soils where leaching or low cation exchange capacity may be a problem.

Foliar Feeding

When deficiency of a minor element is probable and corrective action was not taken in planning the soil fertilization program, foliar feeding can be useful. Such emergencies are infrequent, usually occurring when a vegetable has a high requirement for a micronutrient whose availability is restricted by undesirable soil pH. Routine use of foliar feeding products in vegetable production is not recommended.

Several foliar products are aggressively marketed and advertised. Many contain N, P_2O_5, and K_2O and are basically diluted starter fertilizer materials supplemented with two to five micronutrients. The advertisements for these products try to create the impression that potential profits will be lost if foliar feeding is not a regular part of a fertility program. In reality, indiscriminate use of foliar nutrients and shotgun foliar mixtures, when soil deficiencies do not exist, wastes time and money and decreases profit margins. Once requirements have been met with a good soil fertility program, nutrient loading will not turn a good crop into a super crop. Marketers argue that foliar feeding can provide needed nutrients when environmental stresses limit feeding from the soil. When this happens, however, the stress itself limits growth (not just nutrient uptake), so that any response to foliar nutrients will be limited until the stress is removed, after which time sufficient feeding from the soil can resume.

If you are determined to use foliar nutrients, treat small areas and compare the results to untreated areas. In the unlikely event that a positive result is obtained, check the soil fertility program closely for problems that may need correction.

Plant Analysis

Plant analysis, which reports the concentration of all the essential elements in a growing crop, is now available at a reasonable cost. This information can identify nutrient deficiencies and toxicities from over fertilization. Plant analysis is best used in conjunction with other information such as the soil test, fertilizer program followed, cropping history, and observations on crop development. Once the results are obtained it is generally too late to make corrections on the existing annual crop; for this reason, plant analysis is used primarily in troubleshooting problem areas where insect, weed, or disease pests do not appear to be the culprit.

Despite the above limitation, analyses of healthy crops, along with good record keeping, can provide useful reference points and lead to better interpretive guidelines for local soil and climatic conditions. This information could help in investigating problems that may arise later, and it also provides a benchmark to assess how changes in the fertility program or in any other production practice affect the nutrition of the crop.

Another reason plant analysis has not been more widely used is that the results must be interpreted by an experienced person. Nutrient concentrations vary with the part of the plant sampled and the age of the plant. In addition, factors besides nutrients that limit growth, such as water stress, cool temperatures, and nematodes, can affect the uptake of nutrients and thus the results of the analysis. Moreover, there is often a lack of interpretive guidelines applicable to local growing conditions.

A specific analysis for nitrate-nitrogen in the petioles or leaves of vegetables that have the capacity to accumulate nitrate can provide guidelines on whether to sidedress. This is because the nitrate content of nitrate-accumulating vegetables is closely related to the supply of nitrate in the soil, provided that cool temperatures or low moisture are not inhibiting normal nitrogen uptake by the plant. Interpretive guidelines are currently being established for potatoes based on numerous nitrogen fertilizer rate experiments relating tuber yields to nitrate-nitrogen concentrations of petioles, fresh petiole sap, and whole leaves at specific stages of growth.

Chapter 9

Neem: The Panacea for Organic Farming

Azadirachta indica (Neem) is a tree in the mahogany family Meliaceae. It is one of two species in the genus *Azadirachta*, and is native to India growing in tropical and semi-tropical regions. Its fruits and seeds are the source of neem oil. Other vernacular names include Vembu in (Tamil), Neem (Hindi, Nepali, Urdu), Nim ((Bengali)), Nimm (Punjabi), Arya Veppu (Malayalam), Azad Dirakht (Persian), Nimba (Sanskrit, Oriya), Limdo (Gujarati language) Kadu-Limba (Marathi), Dongoyaro (in some Nigerian languages), Margosa, Neeb (Arabic), Nimtree, Vepu, Vempu, Vepa (Telugu), Bevu (Kannada), *Kodu nimb* (Konkani), Kohomba (Sinhala), Tamar (Burmese), xoan (Vietnamese), (Sdao, Khmer), (Sadaw, Thai), (Hebrew), Paraiso (Spanish), and Indian Lilac (English).

Figure 1: *Azadirachta indica* (Neeem), flowers & leaves

In East Africa it is also known as *Muarubaini* (Swahili), which means *the tree of the 40*, as it is said to treat 40 different diseases, and in Somalia it is known as "Geed Hindi" which means "the Indian tree".

Neem is a fast-growing tree that can reach a height of 15–20 metres (49–66 ft), rarely to 35–40 metres (115–130 ft). It is evergreen, but in severe drought it may shed most or nearly all of its leaves. The branches are wide spread. The fairly dense crown is roundish or oval and may reach the diametre of 15–20 metres (49–66 ft) in old, free-standing specimens.

Leaves

The opposite, pinnate leaves are 20–40 centimetres (7.9–16 in) long, with 20 to 31 medium to dark green leaflets about 3–8 centimetres (1.2–3.1 in) long. The terminal leaflet is often missing. The petioles are short.

Figure 2: Leaves and Fruits of Neem

Flowers

Figure 3: Flowers of Neem

The (white and fragrant) flowers are arranged axillary, normally in more-or-less drooping panicles which are up to 25 centimetres

(9.8 in) long. The inflorescences, which branch up to the third degree, bear from 150 to 250 flowers. An individual flower is 5–6 millimetres (0.20–0.24 in) long and 8–11 millimetres (0.31–0.43 in) wide. Protandrous, bisexual flowers and male flowers exist on the same individual.

Fruit

The fruit is a smooth (glabrous) olive-like drupe which varies in shape from elongate oval to nearly roundish, and when ripe are 1.4–2.8 centimetres (0.55–1.1 in) by 1.0–1.5 centimetres (0.39–0.59 in). The fruit skin (exocarp) is thin and the bitter-sweet pulp (mesocarp) is yellowish-white and very fibrous. The mesocarp is 0.3–0.5 centimetre (0.12–0.20 in) thick. The white, hard inner shell (endocarp) of the fruit encloses one, rarely two or three, elongated seeds (kernels) having a brown seed coat.

The neem tree is very similar in appearance to its relative, the Chinaberry (Melia azedarach). The Chinaberry tree is toxic to most animals, especially to fish, but birds are known to gorge themselves on the Chinaberries, the seeds passing harmlessly through their unique digestive systems.

Ecology

The neem tree is noted for its drought resistance. Normally it thrives in areas with sub-arid to sub-humid conditions, with an annual rainfall 400–1,200 millimetres (16–47 in). It can grow in regions with an annual rainfall below 400 mm, but in such cases it depends largely on ground water levels. Neem can grow in many different types of soil, but it thrives best on well drained deep and sandy soils. It is a typical tropical to subtropical tree and exists at annual mean temperatures between 21–32 °C (70–90 °F). It can tolerate high to very high temperatures and does not tolerate temperature below 4 °C (39 °F). Neem is a life-giving tree, especially for the dry coastal, southern districts of India. It is one of the very few shade-giving trees that thrive in the drought-prone areas. The trees are not at all delicate about the water quality and thrive on the merest trickle of water, whatever the quality. In India it is very common to see neem trees used for shade lining the streets or in most people's back yards. In very dry areas the trees are planted in large tracts of land.

Weed Status

Neem is considered a weed in many areas, including some parts of the Middle East, and most of Sub-Saharan Africa including West

Africa where in Senegal it has been used as a malarial drug and Tanzania and other Indian Ocean states where in Kiswahili it is known as 'the panacea', literally 'the tree that cures forty [diseases]', where ayurvedic uses are practiced.

Ecologically, it survives well in similar environments to its own, for example replacing the babul acacia tree from India with African acacia species.

Chemical Compounds

Indian scientists were the first to bring the plant to the attention of phytopharmacologists. In 1942, while working at the Scientific and Industrial Research Laboratory at Delhi University, British India, three bitter compounds were extracted from neem oil, which were named nimbin, nimbinin, and nimbidin. The seeds contain a complex secondary metabolite azadirachtin.

Uses

In India, the plant is variously known as "Sacred Tree," "Heal All," "Nature's Drugstore," "Village Pharmacy" and "Panacea for all diseases". Products made from neem trees have been used in India for over two millennia for their medicinal properties: neem products are believed to be anthelmintic, antifungal, antidiabetic, antibacterial, antiviral, contraceptive and sedative. Neem products are also used in selectively controlling pests in plants. It is considered a major component in Ayurvedic and Unani medicine and is particularly prescribed for skin disease.

- All parts of the tree are said to have medicinal properties (seeds, leaves, flowers and bark) and are used for preparing many different medical preparations.
- The chemical constituents nimbidin and nimbin have some spermicidal activity
- Neem oil is used for preparing cosmetics (soap, neem shampoo, balms and creams such as Margo soap) and many oral health products.
- Besides its use in traditional Indian medicine, the neem tree is of great importance for its anti-desertification properties and possibly as a good carbon dioxide sink.
- Practitioners of traditional Indian medicine recommend that patients with chicken pox sleep on neem leaves.

- Neem gum is used as a bulking agent and for the preparation of special purpose food.
- Traditionally, slender neem branches have been chewed to clean one's teeth. Neem twigs are still collected and sold in markets for this use, and in India one often sees youngsters in the streets chewing on neem twigs.
- Neem blossoms are used in Andhra Pradesh, Tamil Nadu and Karnataka to prepare Ugadi pachhadi. "Bevina hoovina gojju" (a type of curry prepared with neem blossoms) is common in Karnataka throughout the year. Dried blossoms are used when fresh blossoms are not available. In Tamilnadu, a rasam (veppam poo rasam) made with neem blossoms is a culinary speciality.
- A mixture of neem flowers and bella (jaggery or unrefined brown sugar) is prepared and offered to friends and relatives, symbolic of sweet and bitter events in the upcoming new year.
- Extract of neem leaves is thought to be helpful as malaria prophylaxis despite the fact that no comprehensive clinical studies are yet available. In several cases, private initiatives in Senegal were successful in preventing malaria. However, major NGOs such as USAID are not supposed to use neem tree extracts unless the medical benefit has been proved with clinical studies.

Pest and Disease Control

Neem is a key ingredient in non-pesticidal management (NPM), providing a natural alternative to synthetic pesticides. Neem seeds are ground into a powder that is soaked overnight in water and sprayed onto the crop. To be effective, it is necessary to apply repeatedly, at least every ten days. Neem does not directly kill insects on the crop. It acts as an anti-feedant, repellent, and egg-laying deterrent, protecting the crop from damage. The insects starve and die within a few days. Neem also suppresses the hatching of pest insects from their eggs. Neem cake is often sold as a fertilizer.

As a Vegetable

The tender shoots and flowers of the neem tree are eaten as a vegetable in India. Neem flowers are very popular for their use in Ugadi Pachhadi (soup-like pickle), which is made on Ugadi day in the South Indian States of Andhra Pradesh, Tamilnadu and Karnataka.

A souplike dish called *Veppampoo Rasam* (Tamil) (translated as "neem flower rasam") made of the flower of neem is prepared in Tamil Nadu. In West Bengal, young neem leaves are fried in oil with tiny pieces of eggplant (brinjal). The dish is called *nim begun* and is the first item during a Bengali meal that acts as an appetizer. It is eaten with rice.

Neem is used in parts of mainland Southeast Asia, particularly in Cambodia, Laos (where it is called *kadao*), Thailand (where it is known as *sadao* or *sdao*), Myanmar (where it is known as *tamar*) and Vietnam (where it is known as *s§u ðâu* and is used to cook the salad *gÏi s§u ðâu*). Even lightly cooked, the flavour is quite bitter and the food is not enjoyed by all inhabitants of these nations, though it is believed to be good for one's health. Neem gum is a rich source of protein. In Myanmar, young neem leaves and flower buds are boiled with tamarind fruit to soften its bitterness and eaten as a vegetable. Pickled neem leaves are also eaten with tomato and fish paste sauce in Myanmar.

Patent Controversy

In 1995, the European Patent Office (EPO) granted a patent on an anti-fungal product derived from neem to the US Department of Agriculture and W. R. Grace and Company. The Indian government challenged the patent when it was granted, claiming that the process for which the patent had been granted had actually been in use in India for over 2,000 years. In 2000, the EPO ruled in India's favour but W. R. Grace appealed, claiming that prior art about the product had never been published in a scientific journal. On 8 March 2005, that appeal was lost and the EPO revoked the Neem patent.

Bordeaux Mixture

Bordeaux mixture is a mixture of copper(II) sulfate and slaked lime used as a fungicide in vineyards. It is used mainly to control garden, vineyard, nursery and farm infestations of fungi, primarily downy mildew which can result from infections of *Plasmopara viticola*. It was invented in the Bordeaux region of France, where it is known locally as *Bouillie Bordelaise*. This fungicide has been used for over a century and is still used, although the copper can leach out and pollute streams.

Main Uses

As well as its use to control fungal infection on grape vines the mixture is also widely used to control potato blight, peach leaf curl and apple scab. It is approved for organic use and so is often used by organic gardeners where non-organic gardeners would prefer other controls.

History

In the 19th century there were several outbreaks of vine diseases among the *Vitis vinifera* vines of the classical European wine regions. These outbreaks were caused by pests to which these vines lacked resistance, carried on vines brought to Europe as botanical specimens of American origin. These pests included not only the Great French Wine Blight caused by the aphid *Phylloxera vastatrix*, but also mildew and other diseases caused by fungi.

After the downy mildew had struck, botany professor Pierre-Marie-Alexis Millardet of the University of Bordeaux was studying the disease in vineyards of the Bordeaux region. Millardet then noted that vines closest to the roads did not show mildew, while all other vines were affected. After inquiries, he found out that those vines had been sprayed with a mixture of copper sulfate and lime to deter passersby from eating the grapes, since this treatment was both visible and bitter-tasting. This led Millardet to conduct trials with this treatment. The trials primarily took place in the vineyards of Château Dauzac, where he was assisted by Ernest David, Dauzac's technical director. Millardet published his findings in 1885, and recommended the mixture to combat downy mildew.

In France, the use of Bordeaux mixture has also been known as the Millardet-David treatment.

Mode of Action

Bordeaux mixture achieves its effect by means of the copper ions (Cu^{2+}) of the mixture. These ions affect enzymes in the fungal spores in such a way as to prevent germination. This means that Bordeaux mixture must be used pre-emptively, before the fungal disease has struck.

Preparation

Bordeaux mixture can be prepared using differing proportions of the components. In preparing Bordeaux mixture, the copper sulphate and the lime are dissolved separately in water and then mixed together. Calcium oxide (burnt lime) and calcium hydroxide (hydrated lime) gives the same end result since an excess of water is used in the preparation.

The conventional method of describing the mixture's composition is to give the weight of copper sulphate, the weight of hydrated lime and the volume of water in that order. The percentage of the weight of copper sulphate to the weight of water employed determines the concentration of the mixture. Thus a 1% Bordeaux mixture, which is typical, would have the formula 1:1:100, with the first "1" representing 1 kg copper sulphate (penta-hydrated), the second representing 1 kg hydrated lime, and the 100 representing 100 litres (100 kg) water. As copper sulphate contains 25% copper, the copper content of a 1% Bordeaux mixture would be 0.25%. The quantity of lime used can be lower than that of the copper sulphate. 1 kg copper sulphate actually requires only 0.225 kg of chemically pure hydrated lime to precipitate all the copper. Good proprietary brands of hydrated lime are now freely available but, as even these deteriorate on storage (by absorbing carbon dioxide from the air), a ratio of less than 2:1 is seldom used, which corresponds to a 1:0.5:100 mixture.

Risks

Bordeaux mixture has been found to be harmful to fish, livestock and—due to potential build up of copper in the soil—earthworms.

The chemical began being used by the United Fruit Company throughout Latin America around 1922. The mixture was nicknamed 'perico', or 'parakeet', because it would turn workers completely blue. Many workers would get sick or die of poisoning due to the chemical.

Organic Fertilizer

Naturally occurring organic fertilizers include manure, slurry, worm castings, peat, seaweed, humic acid, and guano. Sewage sludge use in organic agricultural operations in the U.S. has been extremely limited and rare due to USDA prohibition of the practice (due to toxic metal accumulation, among other factors).

Processed organic fertilizers include compost, bloodmeal, bone meal, humic acid, amino acids, and seaweed extracts. Other examples are natural enzyme-digested proteins, fish meal, and feather meal. Decomposing crop residue (green manure) from prior years is another source of fertility.

Discussion of the Term 'Organic'

There used to be a distinction between the term "organic" and the term "pesticide-free". Organic simply dealt with the use of fertilizer types. Once the term "organic" became regulated, many other factors were added. "Pesticide-free" is not at all related to fertilization (plant nutrition), but has become a legal inclusion. In scientific terms, organic compounds must include Carbon.

Natural Sourcing

Animal-sourced Urea and Urea-Formaldehyde (from urine) are suitable for application organic agriculture; however, pure (synthetically produced) urea is not. The common thread that can be seen through these examples is that *organic* agriculture attempts to define itself through minimal processing (e.g., via chemical energy such as petroleum), as well as being naturally occurring or via natural biological processes such as composting.

Cover crops are also grown to enrich soil as a green manure through nitrogen fixation from the atmosphere; as well as phosphorus (through nutrient mobilisation) content of soils.

Powdered limestone, mined rock phosphate, and Chilean saltpeter are *inorganic* chemicals in the technical (organic chemistry) sense of the word but are considered suitable for organic agriculture in *limited amounts*..

Advantages

Although the density of nutrients in organic material is comparatively modest, they have many advantages. The majority of nitrogen-supplying organic fertilizers contain insoluble nitrogen and act as a slow-release fertilizer. By their nature, organic fertilizers increase physical and biological nutrient storage mechanisms in soils, mitigating risks of over-fertilization. Organic fertilizer nutrient content, solubility, and nutrient release rates are typically much lower than mineral (inorganic) fertilizers. A University of North Carolina study found that potential mineralisable nitrogen (PMN) in the soil was 182–285% higher in organic mulched systems than in the synthetics control.

Organic fertilizers also re-emphasize the role of humus and other organic components of soil, which are believed to play several important roles:

- Mobilising existing soil nutrients, so that good growth is achieved with lower nutrient densities while wasting less
- Releasing nutrients at a slower, more consistent rate, helping to avoid a boom-and-bust pattern
- Helping to retain soil moisture, reducing the stress due to temporary moisture stress

- Improving the soil structure
- Helping to prevent topsoil erosion (responsible for desertfication and the Dust bowl.

Organic fertilizers also have the advantage of avoiding certain problems associated with the regular heavy use of artificial fertilizers:

- The necessity of reapplying artificial fertilizers regularly (and perhaps in increasing quantities) to maintain fertility
- Extensive runoff of soluble nitrogen and phosphorus, leading to eutrophication of bodies of water (which causes fish kills)
- Costs are lower for if fertilizer is locally available.

Disadvantages

Organic fertilizers have the following disadvantages:

- As a dilute source of nutrients when compared to inorganic fertilizers, transporting large amount of fertilizer incurs higher costs, especially with slurry and manure.
- The composition of organic fertilizers tends to be more complex and variable than a standardised inorganic product.
- *Improperly*-processed organic fertilizers may contain pathogens from plant or animal matter that are harmful to humans or plants. However, proper composting should remove them.
- More labour is needed to compost organic fertilizer, increasing labour costs. Some of this cost is offset by reduced cash purchase.

Conventional Farming Application

In non-organic farming, a compromise between the use of artificial and organic fertilizers is common, often using inorganic fertilizers supplemented with the application of organics that are readily available such as the return of crop residues or the application of manure.

Vermicompost

Vermicompost is the product or process of composting utilising various species of worms, usually red wigglers, white worms, and earthworms to create a heterogeneous mixture of decomposing vegetable or food waste, bedding materials, and vermicast. Vermicast, similarly known as worm castings, worm humus or worm manure, is the end-product of the breakdown of organic matter by a species of earthworm.

Containing water-soluble nutrients, vermicompost is an excellent, nutrient-rich organic fertilizer and soil conditioner. The process of producing vermicompost is called *vermicomposting*.

Suitable Species

The earthworm species (or composting worms) most often used are Red Wigglers (*Eisenia foetida* or *Eisenia andrei*), but European nightcrawlers (*Eisenia hortensis*) also may be used. Users refer to European nightcrawlers by a variety of other names, including *dendrobaenas*, *dendras*, and Belgian nightcrawlers.

Blueworms (*Perionyx excavatus*) may be used in the tropics. However, *P. excavatus* worms are not suitable for worm compost bins in most of the contiguous United States.

These species commonly are found in organic-rich soils throughout Europe and North America and live in rotting vegetation, compost, and manure piles. They may be an invasive species in some areas. As they are shallow-dwelling and feed on decomposing plant matter in the soil, they adapt easily to living on food or plant waste in the confines of a worm bin.

Composting worms can also be collected from compost and manure piles. These species are not the same worms that are found in ordinary soil or on pavement when the soil is flooded by water.

Large Scale

Large-scale vermicomposting is practised in Canada, Italy, Japan, the Philippines, and the United States . The vermicompost may be used for farming, landscaping, to create compost tea, or for sale. Some of these operations produce worms for bait and/or home vermicomposting.

There are two main methods of large-scale vermiculture. Some systems use a windrow, which consists of bedding materials for the earthworms to live in and acts as a large bin; organic material is added to it. Although the windrow has no physical barriers to prevent worms from escaping, in theory they should not due to an abundance of organic matter for them to feed on. Often windrows are used on a concrete surface to prevent predators from gaining access to the worm population.

Movement of castings through a worm bed.

The second type of large-scale vermicomposting system is the raised bed or flow-through system. Here the worms are fed an inch of "worm chow" across the top of the bed, and an inch of castings are harvested from below by pulling a breaker bar across the large mesh screen which forms the base of the bed.

Because red worms are surface dwellers constantly moving towards the new food source, the flow-through system eliminates the need to separate worms from the castings before packaging. Flow-through systems are well suited to indoor facilities, making them the preferred choice for operations in colder climates.

Small Scale

For vermicomposting at home, a large variety of bins are commercially available, or a variety of adapted containers may be used. They may be made of old plastic containers, wood, Styrofoam, or metal containers. The design of a small bin usually depends on where an individual wishes to store the bin and how they wish to feed the worms.

Some materials are less desirable than others in worm bin construction. Metal containers often conduct heat too readily, are prone to rusting, and may release heavy metals into the vermicompost. Some cedars, Yellow cedar, and Redwood contain resinous oils that may harm worms, although Western Red Cedar has excellent longevity in composting conditions. Hemlock is another inexpensive and fairly rot-resistant wood species that may be used to build worm bins.

Bins need holes or mesh for aeration. Some people add a spout or holes in the bottom for excess liquid to drain into a tray for collection. Worm compost bins made from recycled or semi-recycled plastic are ideal, but require more drainage than wooden ones because they are non-absorbent. However, wooden bins will eventually decay and need to be replaced.

Small-scale vermicomposting is well-suited to turn kitchen waste into high-quality soil amendments, where space is limited. Worms can decompose organic matter without the additional human physical effort (turning the bin) that bin composting requires.

Composting worms which are detritivorous (eaters of trash), such as the red wiggler *Eisenia fetidae*, are epigeic (surface dwellers) together with symbiotic associated microbes are the ideal vectors for decomposing food waste. Common earthworms such as *Lumbricus terrestris* are anecic (deep burrowing) species and hence unsuitable for use in a closed system. Other soil species that contribute include insects, other worms and molds.

Climate and Temperature

The most common worms used in composting systems, redworms (*Eisenia foetida, Eisenia andrei,* and *Lumbricus rubellus*) feed most rapidly at temperatures of 15–25 °C (59-77 °F). They can survive at 10 °C (50 °F). Temperatures above 30 °C (86 °F) may harm them. This temperature range means that indoor vermicomposting with redworms is suitable in all but tropical climates. (Other worms like Perionyx excavatus are suitable for warmer climates.) If a worm bin is kept outside, it should be placed in a sheltered position away from direct sunlight and insulated against frost in winter. It is necessary to monitor the temperatures of large-scale bin systems (which can have high heat-retentive properties), as the feedstocks used can compost, heating up the worm bins as they decay and killing the worms.

Feedstock

There are few food wastes that vermicomposting cannot compost, although meat waste and dairy products are likely to putrefy, and in outdoor bins can attract vermin. Green waste should be added in moderation to avoid heating the bin.

Small-scale or Home Systems

Such systems usually use kitchen and garden waste, using "earthworms and other microorganisms to digest organic wastes, such as kitchen scraps". This includes:

- All fruits and vegetables (including citrus and other "high acid" foods)
- Vegetable and fruit peels and ends
- Coffee grounds and filters
- Tea bags (even those with high tannin levels)
- Grains such as bread, cracker and cereal (including moldy and stale)
- Eggshells (rinsed off)
- Leaves and grass clippings (not sprayed with pesticides).

Large-scale or Commercial

Such vermicomposting systems need reliable sources of large quantities of food. Systems presently operating use:

- Dairy cow or pig manure
- Sewage sludge

- Agricultural waste
- Food processing and grocery waste
- Cafeteria waste
- Grass clippings and wood chips.

Harvesting

Vermicompost is ready for harvest when it contains few-to-no scraps of uneaten food or bedding. There are several methods of harvesting from small-scale systems: "dump and hand sort", "let the worms do the sorting", "alternate containers" and "divide and dump." These differ on the amount of time and labour involved and whether the vermicomposter wants to save as many worms as possible from being trapped in the harvested compost. While harvesting, it's also a good idea to try to pick out as many eggs/cocoons as possible and return them to the bin. Eggs are small, lemon-shaped yellowish objects that can usually be seen pretty easily with the naked eye and picked out.

Properties

Vermicompost has been shown to be richer in many nutrients than compost produced by other composting methods. It also has outperformed a commercial plant medium with nutrients added, but needed adjustment for magnesium and pH.

However, other studies have shown that the effects of home made, backyard, vermicompost compared to municipal compost were lower in terms of soil microbial biomass, soil microbial activity, and yields of a species of ryegrass. Further, one study concluded that the differences between methods of composting were in large part due to the feedstock, and therefore no generalisations can be made between composts made from varying materials.

It is rich in microbial life which converts nutrients already present in the soil into plant-available forms.

Unlike other compost, worm castings also contain worm mucus which helps prevent nutrients from washing away with the first watering and holds moisture better than plain soil.

Bibliography

Ableman, Michael: *From the Good Earth: A Celebration of Growing Food Around the World*, HNA Books, NJ, 1993.

Allen, Patricia: *Food for the Future: Conditions and Contradictions of Sustainability*, Wiley, New York, 1993.

Avery A.: *The Truth About Organic Foods,* Henderson Communications, L.L.C., 2006.

Belasco, Warren James: *Appetite for Change: How the Counterculture Took on the Food Industry*, Cornell University Press, Ithaca, 2007.

Berry, Wendell, and Sierra Club: *The Unsettling of America: Culture & Agriculture*, Sierra Club Books, San Francisco, 1996.

Chadwick, Alan, and Stephen J. Crimi: *Performance in the Garden: A Collection of Talks on Biodynamic French Intensive Horticulture*, Logosophia Press, Mars Hill, NC, 2007.

Conford, Philip: *The Organic Tradition: An Anthology of Writings on Organic Farming, 1900-1950*, Green Books, Bideford, Devon, 1988.

Corum, Vance, Marcie A. Rosenzweig, and Eric Gibson: *The New Farmers' Market: Farm-Fresh Ideas for Producers, Managers & Communities*, New World Publishing; White River Junction, Vt., 2001.

Duram, Leslie A.: *Encyclopedia of Organic, Sustainable, and Local Food,* ABC-CLIO, LLC, Santa Barbara, Calif., 2010.

Fukuoka, Masanobu: *The One-Straw Revolution: An Introduction to Natural Farming*, Rodale Press, Emmaus, 1978.

Gliessman, Stephen R.: *Agroecology: The Ecology of Sustainable Food Systems*, CRC Press, Boca Raton, 2007.

Gussow, Joan Dye: *This Organic Life: Confessions of a Suburban Homesteader,* Chelsea Green Pub., White River Junction, Vt., 2001.

Guthman J.: *Agrarian Dreams: The Parodox of Organic Farming in California*, University of California Press, Berkeley and London, 2004.

Guthman, Julie, and NetLibrary Inc.: *Agrarian Dreams*: *the Paradox of Organic Farming in California*, University of California Press, Berkeley, 2004.

Howard, Albert: *An Agricultural Testament*, Oxford University Press, New York, London, 1943.

Imhoff, Dan, Jo Ann Baumgartner, and Wild Farm Alliance: *Farming and the Fate of Wild Nature: Essays in Conservation-Based Agriculture*, Watershed Media, Healdsburg, Calif., 2006.

Johnson, Wendy: *Gardening at the Dragon's Gate: At Work in the Wild and Cultivated World*, Bantam Books, New York, 2008.

Kingsolver, Barbara, Steven L. Hopp, and Camille Kingsolver: *Animal, Vegetable, Miracle: A Year of Food Life*, HarperCollins Publishers, New York, NY, 2007.

Koepf, Herbert H., Roderick Shouldice, and Walter Goldstein: *The Biodynamic Farm: Agriculture in the Service of the Earth and Humanity*, Anthroposophic Press, Hudson, NY, 1989.

Kristiansen, Paul, Acram Taji, and John P. Reganold: *Organic Agriculture: A Global Perspective*, Comstock Pub. Associates, Ithaca, New York, 2006.

Lampkin & Padel: *The Economics of Organic Farming: An International Perspective*, CAB International, Guildford, 1994.

Lockeretz, William: *Organic Farming: An International History*, Cambridge, MA: CABI, 2007.

Merrill, Richard: *Radical Agriculture*, New York University Press, New York, 1976.

Myers, Adrian: *Organic Futures: The Case for Organic Farming*, Chelsea Green Publishing, White River Junction, VT, 2005.

Nabhan, Gary Paul: *Coming Home to Eat: The Pleasures and Politics of Local Foods*, W.W. Norton, New York, 2002.

Petrini, Carlo: *Slow Food Nation: Why Our Food Should Be Good, Clean, and Fair*, Rizzoli Ex Libris, New York, 2007.

Robinson, Jennifer Meta, and J. A. Hartenfeld: *The Farmers' Market Book: Growing Food, Cultivating Community*, Quarry Books/Indiana University Press, Bloomington, Ind, 2007.

Schilthuis, Willy: *Biodynamic Agriculture*: *Rudolf Steiner's Ideas in Practice*, Anthroposophic Press, Hudson, NY, 1994.

Shiva, Vandana: *Manifestos on the Future of Food & Seed*, South End Press, Cambridge, Mass, 2007.

Swezey, Sean L., and University of California: *Organic Apple Production Manual*, Agriculture and Natural Resources Communication Services-Publications, Oakland, Calif., 2000.

Van den Bosch, Robert: *The Pesticide Conspiracy*, University of California Press, Berkeley, 1989.

Willer, Helga, Minou Yussefi, and Neil Sorensen: *The World of Organic Agriculture: Statistics and Emerging Trends 2008*, Earthscan Publications, London, 2008.

Index

O

P

R

S

❑❑❑